Weather Report: Art and Climate Change

This publication accompanies the exhibition "Weather Report: Art and Climate Change," curated by Lucy R. Lippard and organized by the Boulder Museum of Contemporary Art in collaboration with EcoArts, September 14–December 21, 2007.

1750 13th Street
Boulder, Colorado 80302

bmoca.org

Managing Editor: Kirsten Gerdes

All text accompanying color plates was provided by the artists.

Statements edited by Elizabeth Marglin and Kim Manajek

Biographies edited by Isabella Gonzales

Copy editing by Deirdre Shevlin Bell and Kim Manajek

Production Notes

Design and Production
Vermilion
Boulder, Colorado

Printing
Vision Graphics
Loveland, Colorado. Forest Stewardship Council (FSC)-certified printer. Printed using low-VOC and vegetable-based inks.

Paper
New Leaf Reincarnation Matte 95# Cover and 80# Text contain 100% recycled content, 50% post-consumer waste, and are processed chlorine free.
All New Leaf inventory products are manufactured with 100% renewable energy and are Ancient Forest Friendly.

Thanks to New Leaf Paper, Vision Graphics, and Vermilion for their contributions to this catalog.

All photographs appear courtesy of the artists and/or lenders of the works.

ISBN: 0-9799007-0-0

NEW LEAF PAPER®
ENVIRONMENTAL BENEFITS STATEMENT
of using post-consumer waste fiber vs. virgin fiber

Boulder Museum of Contemporary Art saved the following resources by using New Leaf Reincarnation Matte, manufactured with electricity offset with Green-e® certified renewable energy certificates, 100% recycled fiber and 50% post-consumer waste, and processed chlorine free.

trees	water	energy	solid waste	greenhouse gases
24 fully grown	5,121 gallons	11 million Btu	1,120 pounds	1,893 pounds

Calculations based on research by Environmental Defense and other members of the Paper Task Force.

©2007 New Leaf Paper www.newleafpaper.com

ANCIENT FOREST FRIENDLY™ NEW LEAF PAPER manufactured with wind power

WEATHER REPORT:

ART AND CLIMATE CHANGE

Essays by Lucy R. Lippard, Stephanie Smith, and Andrew C. Revkin

Boulder Museum of Contemporary Art in collaboration with EcoArts

Acknowledgments

I've never bitten off a subject this hard to chew, and neither, I suspect, have most of the artists. Initially I planned to keep *Weather Report* (with apologies to Herbie Hancock, Wayne Shorter, et. al.) local to the Southwest. Then I got carried away, and it became national, then international. Happily, this exhibition was very much a collaborative effort. Thanks to EcoArts founder Marda Kirn, who made me (and many others) do it; Kirsten Gerdes and Joan Markowitz of BMoCA, who were crazy enough to go ahead with it and worked far beyond the call of duty to get it done; and my New Mexico neighbor, artist Isabella Gonzales (with the encouragement of Tom Morin), whose help made it possible. They were all a joy to work with, in part because they all kept their senses of humor.

Thanks also to Stephanie Smith, Amy Lipton, Joy Garnett, Greg Sholette, Mary Jo Aagerstoun, and others who recommended artists to check out for this show, and to Hannah Bird at Cape Farewell, Donna Wingate of Precipice Alliance, Rose Lord at Marian Goodman Gallery, Josie Browne at Max Protetch Gallery, Ronald Feldman Fine Arts, and the Luhring Augustine Gallery. Above all, I'm grateful to all the artists, whose work is so rich and provocative. They have been extraordinarily patient with this small museum's venture into the hot waters of such a huge show.

—Lucy R. Lippard, Curator

An exhibition of this magnitude would not have been possible without the support of a myriad of collaborators. We would like to thank Marda Kirn of EcoArts for her initial vision, for bringing the remarkable Lucy Lippard to the museum, and for the endless hours she spent on the project. During our first meeting, on a cold Sunday morning in January 2006, the vision was made manifest. Little did any of us know how large this project was to become.

Our gratitude goes out to the Andy Warhol Foundation for the Visual Arts for their significant grant. Additional thanks to Bob Morehouse and staff at Vermilion, Jared Polis Foundation, Brad Feld and Amy Batchelor, Audrey and Andy Franklin, Lannan Foundation, Roche Colorado, CHASE, Colorado Council on the Arts, SCFD, 15th Street Studio, and Clean and Green. We extend our appreciation for their continued support to the Boulder Convention and Visitors' Bureau, the City of Boulder, the Boulder Arts Commission, and particularly Donna Gartenmann, whose knowledge and resources were invaluable.

We wish to thank the museum's Board of Directors, Trustees, and volunteers. Their support of this enormous and at times intimidating undertaking never wavered. We thank our staff: Associate Curator Kirsten Gerdes whose tireless and unrelenting communications with the artists and other collaborators made this exhibition a reality; and Travis Allison, Keleigh Asbury, Lindsey Cash, Ashley Mask Harris, Clarissa King, Kim Manajek, Elizabeth Marglin, and Kent Nowlin who each worked diligently to bring this project to fruition.

It is our belief that the role played by artists in interpreting our complicated and tenuous world cannot be overemphasized. We are most indebted to Lucy Lippard and the artists whose visions have shaped this exhibition.

—Penny Barnow and Joan Markowitz, BMoCA Co-Executive Directors

Weather Report: Art and Climate Change has been blessed by a series of incredible yeses. Once upon a late evening I picked up the phone and asked the great Lucy Lippard if she would curate a show on climate change for EcoArts. Wonderfully, she said yes. Later, Lucy and I went to Joan Markowitz at the Boulder Museum of Contemporary Art to see if BMoCA would co-present the show. Just as wonderfully, BMoCA said yes.

Lucy then went to 51 artists and asked them to participate, and they said yes. Many of these artists were interested in working with scientists, renewable energy experts, and others. The artists and I went to those people, and they said yes, too. Grateful thanks to these advisors and co-creators and to those who referred us to them as well.

Special thanks to the funders of who so generously said yes—especially our major contributors, the Compton Foundation, our Anonymous Angel, Ms. Anonymous, Colorado Council on the Arts, the Boulder Arts Commission, and the Schramm Foundation for helping to make this exhibit possible.

Thanks enormously for the behind-the-scenes work of the BMoCA staff, particularly Joan Markowitz, Penny Barnow, and Kirsten Gerdes, who quietly, persistently, and untiringly moved mountains to make it happen, and to the BMoCA Board who trusted and worked and gave so that it could.

EcoArts itself could never have happened without the assistance of an extraordinary group of collaborating organizations and people and our invaluable staff and volunteers. A huge thank you to them for their energy, creativity, and work.

Finally, I would like to thank Lucy Lippard for choosing the artists, and the artists and collaborators for making their pieces, because in the end, none of us helpers would be here without you.

—Marda Kirn, Founder/Director, EcoArts

Weather Report: Expecting the Unexpected

—Lucy R. Lippard

"People in Greenhouses Shouldn't Tell Lies."[1]

I heard somewhere that weather is what you get, and climate is what you expect. In the last few years, it has become abundantly clear that we no longer know what to expect from the climate; in fact—had we only noticed—we haven't been getting the expected for decades. As a firm believer in the ecological truism that everything is connected, I can't imagine an issue more all encompassing—or more challenging. Climate change, following the Gaia hypothesis, touches upon every aspect of life as we know it, even when we don't know it's happening. One small change in the climate triggers microcosmic to macrocosmic chain reactions. Ecosystems, with their infinite sensitivity to change, "describe climate zones by where they grow."[2] Helen and Newton Harrison's *The Mountain in the Greenhouse*—a video of wildflowers being chased up a mountain by warming temperatures—is both metaphorical and emblematic. When the flowers and their dependents reach the top, where will they go?[3]

The very fact of global warming offers startling insights into life on this planet. The existing literature makes it clear that the more research is done, the more the uncertainties increase, but what is rarely disputed is the fact that humans are exacerbating, if not inducing, climate change. It is also possible that even if we do the right thing, we may not be able to turn it around. "The climate we perceive is a metaphor for the sum of weather conditions over a chosen span of time and space," writes Charles Wohlforth.[4]

Having spent a year or more devouring information on global warming,[5] I looked forward to showing off my newfound knowledge, but my task here is not to summarize the science (which Andrew Revkin does with flair) but to cover all too briefly the strategies chosen by the 51 artists in this show, and to imply what that says about the interactions between art, activism, and science (which Stephanie Smith does so well).[6]

One of the principles of this exhibition, agreed upon from the beginning with EcoArts and BMoCA, was to give artists access to scientists working in the fields that they hoped to address. This process was at once fruitful and delicate. (Scientists are determined to remain politically neutral in order to retain their objectivity; artists chafe at constrictions.) Another goal was to put together a show that was beautiful, accessible, and alarming, but not alarmist. Yet once we have heard the cries of the pika (a small, cute, alpine animal in line to be one of the next species to reach extinction due to climate change, joining the already-lost Golden Toad, Staghorn Coral, and others), as broadcast by Brian Collier in *The Pika Alarm,* we have no choice but to empathize directly with the fate of "Nature," and remember that we too are part of it.

Working with a small museum, and given my own preoccupation with activist and public arts, there were certain choices that had to be made. We knew we could not possibly cover all the bases suggested by climate change. How didactic did we want to be? How aesthetic? How focused? How ambiguous? An informative photography show on the visible effects of global warming would not be hard to put together.[7] A more aesthetically oriented show would incorporate painting and sculpture.[8] I decided to focus on conceptual and site-specific work that seemed better able to cope with the vast amount of information available. We had a large pool of artists to choose from, given resources like Colorado's own EcoArts, the similarly named artists' internetwork, eco-art, the Green Museum, and ecoartspace.[9]

The artists I finally asked to participate (29 women, 12 men, 10 collaborations) had already addressed the subject of climate change or were wholly involved in environmental issues and had made related work. They are important because they reach out to their audiences, they are willing to work in the world, to make art in a global arena, and because they are knowledgeable or curious enough to enter into dialogues with scientists and/or community. It has been exciting to watch them wrestle with this gigantic problem, trying to find ways to follow the debates and communicate visually what we can barely comprehend, even as it is oversimplified in the mass media. Each artist has been moved by a different aspect of the coming changes, ranging from renewable energy sources, sustainable building, food production, habitat restoration, suburban sprawl, governmental responsibility, wildlife migration, water shortages, desertification, and biological mutation, to the economic ramifications of our wasteful society, and on and on. There is no part of human life that will not be touched by climate change over the next millennium—should we last that long.

As Elizabeth Kolbert observes in her book *Field Notes from a Catastrophe,* which has been called "the *Silent Spring* of our time,"[10] the effects of global warming are most visible in sparsely populated areas where "nature" herself is most visible. Boulder, as a city in an arid state with ready access to mountains and grasslands and an impressive record of open space and conservation initiatives, is just the right location for this event. Wildlife migration, growing patterns, rising temperatures, and melting glaciers are all more obvious here than in urban areas that will be equally affected. Yet while Boulder is among the relatively few towns in the United States to have signed a resolution to meet the standards of the Kyoto Protocol and is a hotbed of national and academic climate change institutions (many of which have collaborated with us), the city is far less accustomed to issue-oriented public art. *Weather Report* will deluge the populace with artwork outdoors, around town, at the National Center for Atmospheric Research (NCAR), and at the University of Colorado's Norlin Library and ATLAS building.

The days when artists accepted "cultural confinement," simply making objects to be exhibited in designated spaces, are long gone. Public works are the core of this exhibition. (Since I'm writing this in July, I've not yet seen any of them!) Public art is hard to conceive and execute and vulnerable to vilification and vandalism, since broader audiences who come upon these works unaware are not usually accustomed to the vagaries of the contemporary avant-garde. At the same time, the element of surprise can be invaluable when dealing with a subject that is both familiar and unknown. We will be taking a poll in the museum to see what the most popular works are, how effective and/or moving people find them, and which works are most energizing toward action and change.

In a recent issue of *The Nation* devoted to "the climate crisis," Lawrence Wechsler's article on "artistic responses to global warming" appeared 10th of 10 articles. It is not unusual for art to come in last where "real world" issues are involved. Wechsler pretty much dismisses artists who address the issue "head-on, often in the form of exhibitions with massive science-fair-like wall postings, lots and lots of words, all quite earnest though in the end not terribly satisfying as art" or as agitprop.[11] I do not agree with this oversimplification and have tried to put together a show so varied that it cannot be dismissed as merely art, merely science, or merely agitprop.

There is no reason to exaggerate the elusive power of art. Artists cannot change the world ... alone. But when they make a concerted effort, they collaborate with life itself. Working with and between other disciplines and audiences, and given the chance to be seriously considered outside the rather narrow world of art, they can offer visual jolts and subtle nudges to conventional knowledge. The popular image of artists as renegades frees them to imagine situations and outcomes beyond the boxes. At best they can make the hot breath of climate change both vivid and immediate to this visually oriented society, and they can inform us in the process. They can also deconstruct the ways we are manipulated by the powers that be and help open our eyes to what we must do to resist and survive.

Newton Harrison and Helen Mayer Harrison are North American pioneers of what is now called ecological or "eco" art, having worked since the 1970s with scientists to envision alternative systems for entire countries, watersheds—even continents.[12] As they unearth the stories buried in every map, they construct new scenarios with which to imagine a different future. Patricia Johanson's "buildings that grow and change," forming alliances with living systems in order to "subsume the hubris of ego and waste," were conceived in the 1960s, when she was commissioned by *House and Garden* to design a garden. Although unpublished, this engendered some 150 sketches and set her on a lifelong engagement with public art, energy conservation, and sustainability. Working around the world, she has blurred the boundaries between the habitats, patterns, and processes that "enmesh human creation within the larger patterns and purposes of the natural world."[13]

Another early contributor to art about global warming was Agnes Denes, whose many years of visionary environmental projects and propositions are devoted to creative problem-solving. Her 1980 *Pascal's Perfect Probability Pyramid & the People Paradox—The Predicament (PPPPPPP)* depicts more than 16,000 tiny people forming the structure of a pyramid—a metaphor for interdependence, or a "society of visual mathematics." Her *Tree Mountain, A Living Time Capsule* reclaimed a 29 meter–high, human-made mountain in Finland, where 11,000 trees were planted by as many people as a national commitment to the 1992 Rio de Janeiro Earth Summit.

A prescient urban approach to climate change was *Rapid: Post Petroleum Gas Stations—Launching a Brand,* a 1999 indoor/outdoor show anticipating alternative fuel stations, at Nikolai Fine Arts in New York, which happened to be next to an Exxon station. Rapid Response (two artists, two environmental activists) placed a light box sign reading "GLOBAL WARMING," next to and approximating the gas station's signage; the logo style of British Petroleum (now, significantly, "Beyond Petroleum") was appropriated for a sign reading "GREENHOUSE EFFECT." Indoors, minimalist black "pumps" bore detailed information on renewable energy. Above all, Rapid Response tackled practical issues by designing in detail a suspended fuel tank and a Rapid Fuel Station.[14]

Virtually none of us is innocent of overproducing the carbon emissions generated by cars, planes, buildings, and appliances. Taking responsibility for our personal behavior is the theme of Sherry Wiggins' interactive *Carbon Portraits*—profiles of local citizens with varying lifestyles and income levels. The profiles are exhibited in public spaces, where viewers can identify, or not, and apply these analytical tools to their own lives ... or not.[15] In several instances, the artists have attempted to inform the public about how to do the right thing. *Weather Report*'s wide range of temporary works in public spaces include Lynne Hull's *Cooler Choice* stickers, highlighting environmentally friendly choices for consumers. Hull is best known for her "trans-species art," in which she provides amenities for creatures whose habitats are disappearing. Included in the exhibit are a series of "wildlife warning" signs, showing semi-erased animals threatened by lack of water and global warming.

Isabella Gonzales's *2¢ Worth ... Your Participation Is Requested* solicits public action about the thinning of the ozone layer (roughly the thickness of two pennies). Her penny jars (one of which is set up in a local credit union), are nostalgic reminders of some of our childhoods, now eliciting public contemplation about the fragility of the future. The next generations are also the concern of Mierle Laderman Ukeles, famous for her decades of "maintenance art" and work on the production and management of urban solid waste. She will distribute a questionnaire for children ages 7 and up, the drawn or written answers to which will be exhibited in public places. Buster Simpson's *Stretched Shirts*—a photographic scroll documenting a series of "agitprops" about global warming, was executed with a group of high school students, scientists, and artists for a week in Portland, Oregon in 2002. With the light shining through them, the connected white shirts evoke flags or clean laundry (integrity)—vulnerable stand-ins for human resistance and collective effort—all a "stretch" in our complacent society. Simpson is also working on a master plan for the coming Winter Olympics in Vancouver that will incorporate climate change awareness.

One lesson easily taken from this work is the importance of specificity and place-based research, both for scientific data and outreaching art. Eve Andrée Laramée, who has been making videos and installations on the environmental future of the West, is designing a newspaper insert/leaflet to provide scientific information on the changing perceptions of desertification and its relevance to the Boulder area.[16] Bill Gilbert, working with biologist Bruce Milne and artist Erica Osborne, proposes *Bough*—a fractal-determined design for Boulder's mountain/prairie ecotone, utilizing plant species capable of adapting to rising temperatures (in other words, plants that will not fall off the top of the mountain).[17] The biological future is also the subject of Rebecca DiDomenico's *Intentional Mutations*—a shapeshifting menagerie, guided by the artist's scientific collaborators, of creatures that could evolve as wildlife gradually adapts to changing climates. Nature, like artists, she points out, will experiment with evolutionary possibilities, offering "a giant testing ground" for the success and failure of species.

Marguerite Kahrl's *Homeland Insecurity* is a set of potted industrial hemp plants (made of fabric) embedded with subliminal messages that convey factual data about this beneficial crop, countering the bad rep of its cannabis cousin, which has retarded hemp's environmental and economic potential.[18] Janet Koenig and Greg Sholette cite statistics about the 750 million personal computers in use around the world at this very minute. Their plastic-wrapped computer discards, which, improperly recycled, contribute carbons and hazardous chemicals to global warming and to global ill health, are made into monumental columns, doubling as kiosks for posters bearing information on the subject.

The first initiative of the global warming–focused arts organization Precipice Alliance was Mary Ellen Carroll's *indestructible language* (2006)—a 900-foot-long neon text installation across the facade of an old factory building in Jersey City, visible from the Newark Airport: "It is green thinks nature even in the dark." I read this as a hopeful message about nature's capacity for renewal even as we remain stubbornly in the dark about what we must do to maintain the green. Precipice Alliance's cofounder, photographer Joel Sternfeld, contributes to *Weather Report* a group of snapshot-portraits of the participants in the 11th United Nations conference on climate change, in Montreal in 2005. He lets us see the human faces and hear the words at a watershed moment—when the world ignored the United States' determination to undermine the Kyoto Protocol and agreed to go forward and seek solutions.[19]

Several of the public works in *Weather Report* are performance-related. In the 2006 4th of July parade in the small town of Whitefield, Maine, activist artist Natasha Mayers and friends spoofed life in a transformed northern clime (skiing in a bathing suit, SPF 3000, pond hockey, "Snow Plow Drivers for the Kyoto Protocol," L.L.Bean hot-weather fashions carved out of old parkas, and so forth), hilariously bringing the issues of global warming into a local context. Futurefarmers (Amy Franceschini and Michael Swaine), collaborating with microbiologist Jonathan Meuser and Boulder's Masala community, will present an outdoor theatre piece, a kind of fairy tale about a gardener, a community planner, and a scientist who escape to Greenland and set about to solve the world's problems. Bobbe Besold is organizing the *Thirteenth Tipping Point,* a Day of the Dead costumed procession/performance community event to take place on October 29th, dramatically presenting "death and transformation, denial and action".

Basia Irland (who was raised in Boulder) has developed another chapter in her ongoing *A Gathering of Waters* (also executed along the Rio Grande, in Canada and England).[20] This broadly collaborative project connects people along 47 miles of Boulder Creek, drawing attention to the diminished Arapaho Glacier, the source of Boulder's water, which has dropped over 100 feet since 1960. The intermountain West's concern with water is also the subject of Chrissie Orr's vast, lyrical *Earth Drawing,* a map of the shifting watershed, based on climate change modeling from Hydrosphere. "Being in the community, not in a studio, is activism in and of itself," says Orr, who teaches Land Art and Community Activism at Santa Fe's Ecoversity.[21] Jane McMahan, working with glaciologist Tad Pfieffer (a photographer himself), gives form to Boulder's water supply with a block of glacial ice. A "kind of altar," placed near Boulder Creek, it is paradoxically kept "on life support" by energy from solar panels, even as the sun melts the glacier. Israeli artist Shai Zakai documents her projects based on lessons learned from ancient times about adaptation and mitigation (rainwater stored in pebble fields, a technique also used by the Pueblo peoples of the Southwest), a "library" of seeds, and erosion through an exploration of roots, actual and mythical.

While potential lack of water is the subject of several works, Mary Miss (thinking of the water marks on New Orleans buildings after Hurricane Katrina) asks passersby to imagine Boulder's next major flood. Working with Denver hydrologist Steve Blake, she indicates the potential highwater marks of a 100-year flood on buildings and trees around town with large blue dots—"prompts" or "acupuncture" with which to prick the viewers' consciousness as they "connect the dots," and look at our "failure to combine smart water management with smart land use," as Peter Gleick has noted. "We have economic incentives and subsidies that lead people to make bad decisions about placing themselves in harm's way, and even in the absence of climate change we could reduce our flood damage and flood risk and flood deaths enormously by changing the way we build things ... if we were smarter we would rethink these decisions."[22]

Extreme weather events like floods and storms are of course a major source of anxiety. It has been suggested that hurricanes can be regarded as nature's way of driving human infrastructure away from the coastlines to preserve essential ecologies,[23] a possibility illuminated by the Center for Land Use Interpretation's photographs of a barrier island on the Gulf Coast, where vacation homes on the devastated beach have been surreally raised on higher and higher stilts, representing humans' stubborn denials of disaster as well as "the outer limits of habitability on the nation's disintegrating margins." CLUI's deadpan approach to dramatic issues reflects the public indifference to such obvious threats.[24]

For most of us, global warming remains an abstraction that has not yet entered our own lived experiences, despite the innumerable concrete effects already changing our lives offstage. Though we no longer see photography as "truth," it still conveys first-hand experience like no other medium, and has proved effective in bringing distant places closer to oblivious audiences. When Subhankar Banerjee's striking images of the Arctic National Wildlife Reserve (ANWR) were shown at the Smithsonian during Congressional debates on drilling in the refuge, the show was kicked downstairs and its explanatory captions were censored.[25] Decontextualized, even the most beautiful pictures can lose their teeth. ANWR, and the Arctic in general, have been ominously heralded as America's "Last

Frontier." In Banerjee's expressive work, vehicle tracks across the vast tundra become as menacing as a terrorist alert; a whaler's wooden coffin washed out of a beach bank, the bones of humans, the bones of caribou starved of their usual diets, are contrasted with the sustainable hunting of Gwich'in people, who have lived interdependently with Arctic wildlife for millennia.

Like Banerjee's thousands of caribou streaming across their homeland, Chris Jordan drives home his point through mind-boggling multiplicity. His digital photograph of Alaska's immense Mt. Denali is made up of the word "denial" and 24,000 logos representing six weeks of sales of the GMC Yukon Denali SUV; his eight million toothpicks match the number of trees harvested each month to make mail order catalogues; 11,000 jet trails represent eight hours' worth of commercial flights in the United States. Steve Deo (Creek/Euchee) has developed a similar strategy for his social sculptures (such as a globe made up of thousands of plastic toy soldiers). His four life-size bare trees (one of which is appropriately shown in the university's library)—*tree of knowledge, tree of life, heartwood,* and *tree of origin*—are made from thousands of pages of encyclopedias, demonstrating that for all our knowledge, we are still destroying our most precious resources. Without the trees that recycle carbon detritus (and sometimes, perversely, emit it), we cannot survive.[26]

In the Arctic, which has become the canary in the coal mine for global warming, indigenous knowledge is finally being tapped. As one Athabascan elder put it, "You young people, the weather's not going to be at the right place at the right time." Richard Glenn, an Inupiat whaler and ice scientist, mused that scientists know "a collection of facts about ice; Eskimos know ice itself."[27] On our communal behalf, several artists and groups have, like Banerjee, experienced the Arctic's lessons first-hand. British photographer David Buckland founded Cape Farewell in 2001 to bring artists, scientists, and educators together in a film, an exhibition, a book, and a series of voyages in a century-old schooner to an Arctic island, where artists from various fields, including Rachel Whiteread and Anthony Gormley, have made pieces on site.[28]

Lillian Ball's *66 Degrees, 32 Minutes, 50 Years* is the projected history of an ice cap from 1990 to 2040, sparked by an invitation to carve ice in Lapland, above the Arctic Circle, where tree lines are rising and the reindeer herds of the indigenous Sami are starving for lack of lichen. A bowl of melted ice from the site is colored by a map projection from above. Iñigo Manglano-Ovalle's *Iceberg (r11i01)*—a 25-foot aluminum skeleton, like the ghost of a giant creature—was constructed from concrete scientific data by scanning with radar and sonar an existing iceberg drifting in the Labrador Sea.

Global warming's effect on the Antarctic is a bit more subtle, not yet poised as clearly on the edge of the abyss as the Arctic. The otherworldly vistas of the southern polar regions lend themselves to fantasy. French artist Pierre Huyghe's *A Journey That Wasn't*—a 2005 performance at New York's Wollman Ice Rink (standing in for Antarctica)—took as its premise a voyage to the receding Antarctic ice shelf in search of newly revealed islands and a mythical white animal, finally encountered as "a white shadow."[29] Joan Myers began photographing in Antarctica in 2001 and later spent four months there—her acutely trained eye for landscape challenged by a place that looks like no other, which most of us will never see, where change is surfacing from a deep burial.[30] Xavier Cortada created a series of extensive installations on his Antarctic visit, focusing on the crucial element of time. The Florida native cast an ice replica of a Mangrove seedling that is very slowly making its way over 150,000 miles to the coastline. In contrast, he marked the mere 50-year history of human presence at the South Pole with chronological flags stretching across the frozen land.

Several of these artists have been working for years with scientists, and their task is to translate their considerable accumulated knowledge into something the rest of us can comprehend. Judit Hersko's *Seven Days of Dissolution,* part of a larger piece called *Shifting Baselines,* is "an analogy for social processes that include gradual shifts to points of reference previously unacceptable." In seven glowing units, it follows, in flashes, the disintegration of a heart and lungs made from shell suspended in water, connected to a tank of "carbon dioxide."

Based on research in collaboration with oceanographer Victoria Fabry, Hersko chronicles the acidifying effect of increasing carbon dioxide emissions on the oceans, which could eventually destroy much marine life. Like the voyage of Cortada's mangrove, these changes are so slow and almost imperceptible that they go unnoticed or are accepted as the natural order until illuminated by science and art.

Andrea Polli makes audio/visual art that radiates from the vortex of art and science into other fields through interactive digital media. Her *N.* (with Joe Gilmore) is a "sonification" of Arctic weather data from the North Pole, 2003 to 2006, modeled by meteorologist Dr. Patrick Market, resulting in a circular projection that makes distance visible. Like much of her work, such as *Airlight* (referring to a visible white smog caused by light on fine airborne dust particles), it offers new readings of storms and climate through sound.[31]

"What the world needs is a good housekeeper," says Aviva Rahmani, a feminist and eco-artist based on a Maine island. As part of her long-term project, *The Cities and Oceans of If,* Rahmani has generated conversations and digital prints with Dr. Jim White of the Institute of Arctic and Alpine Research, "to clarify the complex relationships between natural resources and built infrastructures" that can be directly correlated with global warming. As they hone in on restoration of degraded environments through "trigger points" in the landscape, they explore ways for artists and scientists to work together internationally. Ellen Levy's signature tall panels, painted over digital imagery (in one case a 1971 *Whole Earth Catalogue*), offer vertiginous perspectives on tightly ordered images and information. Her sources are an emblematic vortex of pragmatism and imagination—plans from the US Patent and Trademark Office, in this case those dealing with aspects of global warming—with a subtext about corporate attempts to "own" existing life forms.[32]

Buildings account for over one-third of our energy use,[33] and sprawl—the ubiquitous phenomenon of subdivisions gobbling up the landscape—destroys agricultural land and open space, wasting precious resources with its labyrinth of roads and infrastructure. Ruth Wallen's *Preserving Paradise* explores through a set of postcards the construction of a California suburb, raising issues of tourism, placelessness, and the transmission of information in visual fragments. Kim Abeles has focused for decades on the urban environmental experience. Making clear where the buck has stopped in the industrial history leading up to pollution and climate change, her 1992 *Presidential Commemorative Smog Plates* (portraits of and quotations from US presidents from McKinley to Bush) are etched by exposure to particulate matter in the air of Los Angeles.

Slovenian artist Marjetica Potrč's spontaneous drawings of inventive solutions to aspects of climate change in specific cities (Pittsburgh, Boulder) are a continuation of her work in beleaguered urban communities around the world (Caracas, Bologna, Belfast, and Ljubljana, among others). As she discovers prototypes of innovative products to improve the lives of the resourceful poor, she considers herself a "storyteller who builds stories with architectural material."[34]

As certain interest groups push for ethanol and hydrogen fuels, despite the fact that they will consume more land and fuel in their construction than renewable sources like wind, solar, and tidal energy, artists explore alternative materials. Kristine Smock's wild "goddess of energy" demands attention as she literally embodies renewable energy sources with her biomass skirt, solar bosom, and windblown hair. Erik Sommerfeld's students at the University of Colorado construct handsome furniture on a sustainable, biological model in his "Symbiotic Design" class. Gayle Crites' delicate drawings commemorate the scientific search for bio-fuel from algae (as well as the beleaguered polar bear's loss of winter sea ice and an abstract vision of the Kyoto Protocol). Demonstrating wind power in aesthetic form are both Patrick Marold's monumental *Windmill Project* in Vail, Colorado, which converts wind into light, and *Coal Warm Memorial*—an installation at NCAR by Airworks (Melanie Walker and George Peters), representing the desire for renewable energy with a tunnel of sound and lights, anchored by rocks ("coal").

Learning Site is one of the now ubiquitous and often youthful collectives working on social and ecological issues across national and continental boundaries. Here they represent the tip of the iceberg of what is being done in Europe and elsewhere about the "Greenhouse Effect." Members Cecilia Wendt and Rikke Luther's "Learning Posters" illustrate projects they have constructed in Europe. In Boulder they are distributing images of a solar module for a greenhouse that is a new addition to a floating house and island they built in 2000 in Copenhagen—hoping to trigger a discussion on "how political consensus is constructed."

Humor and satire are major components of much contemporary activist art. Canadian artist Iain Baxter& has been looking at the environmental ramifications of business as usual for over forty years, initially as the N.E. Thing Co. His *Killer Still Life #5* (a shelf full of household pesticides) and *Animal Preserve #8* (a medicine cabinet of stuffed animals in jars) make their points with humor rather than information overkill.[35] Beverly Naidus's digital prints, *Brought to You by "The Men Who See Beyond the Future,"* satirize corporate advertising with the subversive art practice of "culture jamming," at which she is a longtime expert.

And finally, The Yes Men, masters of satirical performance, have infiltrated some of the globe's most prestigious business/scientific meetings, offering reasonable solutions for the world's problems that seem preposterous only because of the lunatic context in which nations and businesses operate. For instance, as "representatives of Halliburton" at a corporate retreat, they promoted the "Survivaball" suit as a necessary (if expensive, at $100 million) protection against extreme global warming. Their *Ice Age* video takes George W. Bush over the thin ice he has asked for by denying climate change, and most recently, posing as representatives of Exxon at Canada's largest oil conference, The Yes Men modestly proposed the use of victims of climate change as the source for a new alternative fuel to be called Vivoleum.[36]

What next? Will the children of Boulder and beyond take heed from Ukeles' messages? Will Gonzales's pennies offset the US government's reluctance to curb emissions? Will Denes's forests be reduced to Jordan's toothpicks, flooding the planet with carbon as oceans rise and deserts grow? Will the Harrisons' wildflowers find a new home? Will nature heal herself or drag us along a different path? As the global warnings proliferate, we are seeing only the tips of the melting icebergs. But at least we are finally seeing them, and it is the artist's job to teach us how to see.

Lucy R. Lippard is a writer, activist, and author of twenty books on contemporary art and cultural criticism. She has curated over 50 exhibitions. The most recent books are The Lure of the Local: Senses of Place in a Multicentered Society *and* On the Beaten Track: Tourism, Art and Place. *She lives off the grid in Galisteo, New Mexico.*

Endnotes

1 Chapter heading in Tim Flannery, *The Weather Makers: How Man is Changing the Climate and What It Means for Life on Earth,* New York: Atlantic Monthly Press, 2005, 239.

2 Charles Wohlforth, *The Whale and the Supercomputer: On the Northern Front of Climate Change,* New York: North Point Press (Farrar, Straus & Giroux), 2004, 179.

3 Or in a macrocosmic view: "With world ecosystems migrating toward the poles, the ecosystem nearest to the pole would be pushed right off the planet." *(Ibid. 263)*

4 Wohlforth, *op cit,* 150.

5 For a general guide, see Alex Steffen, ed., *World Changing: A User's Guide for the 21st Century,* New York: Abrams, 2006. For background, see J. Madeleine Nash, *El Niño: Unlocking the Secrets of the Master Weather-Maker,* New York: Warner Books, 2002.

6 Stephanie Smith's catalogue *Beyond Green: Toward a Sustainable Art,* Chicago: Smart Museum of Art, 2005, was an inspiration, as was Andrew Revkin's *The Burning Season,* New York, Penguin Books, 1994, and *The North Pole Was Here: Puzzles and Perils at the Top of the World,* Boston: Kingfisher, 2006.

7 Like *NORTHSOUTHEASTWEST* sponsored by the British Council and the Climate Group, which I saw in Venice in 2007.

8 When I forewent painting and sculpture in favor of conceptual work, I found it hard to give up Joyce Kozloff, Christy Rupp, Joy Garnett, Alexis Rockman, and others. Another task was to avoid the temptation to talk about the work of still other artists, who, for various reasons and circumstances, could not be included, such as Olafur Eliasson, Dan Peterman, Ocean/Earth, or the Institute for Figuring's knitted coral reef, featured at the Chicago Humanities Festival in October 2007.

9 Paving the way for all of these was Barbara C. Matilsky's pioneering exhibition/book, *Fragile Ecologies: Contemporary Artists' Interpretations and Solutions,* New York: Rizzoli, 1992.

10 Elizabeth Kolbert, *Field Notes from a Catastrophe,* London: Bloomsbury, 2007.

11 Lawrence Wechsler, "Fevered Imagination," *The Nation,* May 7, 2007, 38.

12 There are many catalogues of their work; see for instance, *Peninsula Europe: The High Grounds,* Newton and Helen Harrison: Del Mar, CA., 2001.

13 Patrician Johanson, "Designing Beyond the Visual: Life, Death, and Tradeoffs in the Garden of Art," talk given to the College Art Association, 2000. See also her booklet *Art and Survival: Creative Solutions to Environmental Problems,* North Vancouver: Gallerie Publications, 1992.

14 See Cathy Lebowitz, "Honey, let's sell the gas guzzler and buy a used Dodge Intrepid," *Art in America,* July 2000, 101,119.

15 Consider the shameful irony of flying around the country to talk about climate change; at least I don't have a private jet like Al Gore, whose film *An Inconvenient Truth* has nevertheless provided a sterling wakeup call.

16 Tim Flannery distinguishes between desertification (a permanent shift to a dry climate; the rains simply do not return) and drought, which is by nature transient. He also discusses the potential for "polar deserts" in the Arctic. (*op.cit.,* 124, 103.)

17 Actualization of this projected outdoor piece turned out to be too much for our resources, but the team seems determined to realize it in the future. Also shown are Gilbert's walking/mapping pieces in homage to John Wesley Powell's pioneering proposals for the arid lands of the West in the 1870s, which have led to contemporary suggestions for bioregional re-organization of the West, according to watersheds, rather than the current artificial state boundaries.

18 The bad reputation of its cousin is the least of the problems. Industrial hemp has over 40,000 uses and could replace any number of petroleum-based products. The real reason it has been outlawed by the US is protection of the industries that would be replaced by hemp products.

19 These works are drawn from Sternfeld's book, *When It Changed,* London: Steidl, 2007.

20 See Irland's forthcoming book *Water Library,* Albuquerque: University of New Mexico Press, 2007.

21 "Ecoversity's Chrissie Orr: A Life in Art, Activism and Community," *The Art of Permaculture,* Nov. 2006, 23.

22 "Interview with Dr. Peter H. Gleick," *American Rivers,* winter/spring, 2007, 8, 9.

23 Dr. Michele Dionne, cited in correspondence from Aviva Rahmani to Jim White, March 27, 2007.

24 See Matthew Coolidge and Sarah Simons, eds., *Overlook: Exploring the Internal Fringes of America with the Center for Land Use Interpretation,* New York: Metropolis Books/DAP, 2006.

25 See Finis Dunaway, "Reframing the Last Frontier: Subhankar Banerjee and the Visual Politics of the Arctic National Wildlife Refuge," *American Quarterly,* v.58, no.1, March 2006, 159-189.

26 Two other Native American artists who have commented on climate change are Cheyenne Arapaho Edgar Heap of Birds in a 1990 text piece "For the Rainforest": (THE LIZARD FEEDS US/BURNING DEBT/BURNING BURGERS/ FOOD FOR THE BRICK PRICK) and Coast Salish Lawrence Yuxweluptun's painting *Red Man Watching White Man Trying to Fix Hole in Sky* (1990).

27 Catherine Attla (Athabascan), and indirect quote from Richard Glenn, Inupiat whaler and ice scientist, in Wohlforth, *op cit.* 255, 128.

28 The book is *Burning Ice: Art & Climate Change,* Layerthorpe: York Publishing Services LTD, 2007.

29 The journey's "diary" is one component in a strange and complex multipartite piece called *Celebration Park.* See Nancy Princenthal, "Reasons to be Happy: Huyghe's 'Celebration Park,'" *Art in America,* Sept. 2006, 125–28.

30 Melting ice recently unearthed giant ferns there. See also Joan Myers, *Wondrous Cold: An Antarctic Journey,* Washington DC: Smithsonian Books, 2006.

31 Polli's creative process is evident in her blog at www.andreapolli.com.

32 See the essay by Laurence Hegarty in *Ellen K. Levy: Evolution n,* Middletown, CT: Wesleyan University, 2006.

33 See "What Can We Do?", a symposium on sustainable design and climate change, noting that a 2001 International Panel on Climate Change determined that buildings and their appliances are responsible for more CO_2 emissions than transportation worldwide (*Harvard Design Magazine,* Spring/Summer 2003, 47–61).

34 See Eleanor Heartney, "A House of Parts," *Art in America,* May, 2004, 140–43; and Sabir Khan, "Marjetica Potrč's Shanties to Go," *Art Papers,* July–Aug. 2004, 22–27.

35 See *Iain Baxter&: Products, Place, Phenomenon,* Windsor: Art Gallery of Windsor, 1996.

36 Conference participants solemnly lit "Vivoleum commemorative candles," purportedly made from the remains of an Exxon janitor who died as the result of a toxic spill, before realizing that they'd been had.

Weather Systems: Questions About Art and Climate Change

—Stephanie Smith

quiet
quiet down she said
speaking to the back of his head
on the edge of her bed
i can see your blood flow
your cells grow

hold still a while
don't spill the wine
i can see it all from here
i can see
i can see
weather systems of the world

and every time you turn the soil
another cloud begins to boil

—Andrew Bird
Excerpted lyrics from the song "Weather Systems," 2003

I have been listening to a lot of Andrew Bird's music lately so the title of this exhibition sent me reaching for his 2003 song "Weather Systems." Stretched across six-and-a-half minutes of complex rhythms and shimmering violin, the song's taut lyrics sketch a set of relationships: she and he, they and world, micro and macro, part and whole. Although by no means *about* climate change—its weather systems are internal rather than external—the song's imagery of bittersweet interdependence has stayed in mind as I have been thinking about this essay. Our actions can wreak havoc and set the clouds to boiling: true enough. But what do we do once metaphoric or real weather systems have gone awry? What can we do, other than retreat to the bedroom, drink in hand, to try to divine the state of things by studying the whorls of hair on the back of a head?

Weather Report raises this question within the realm of visual art. Curator Lucy Lippard has asked how artists are responding to climate change, an inquiry that has yielded this wonderful assemblage. The exhibition includes a staggering range of tactics and approaches, many of which could be framed as part of a broader area of practice that I have described elsewhere as sustainable art.[1] In distinction from earlier generations as well as current practitioners of eco- or environmental art, sustainable art reaches beyond the intersection of environment and aesthetics to attend to ethical issues like social justice and intergenerational equity, topics that are also central to debates on climate change.[2] Recently we have seen a dramatic rise in interest in this matter, with increasing numbers of exhibitions, publications, and conferences dealing with environmental and sustainable art. Climate change is also beginning to get more attention from within cultural arenas.[3]

Particularly in the United States, this artistic surge parallels a rapid rise in public attention to climate change, which in turn builds on an upsurge of interest in "green" issues that has been percolating in the country since about 2000. We seem to have reached a media tipping point over the past year. Al Gore's 2006 documentary *An Inconvenient Truth*, for instance, famously helped catalyze widespread attention to climate change. Whether or not you agree with his arguments or style, Gore's film succeeded in translating serious scientific and policy research into accessible terms, in offering practical if limited answers to the question, "But what can I do about it?" and in using commercial infrastructure—the apparatus of major cinematic release—to reach wide audiences. In early 2007, the release of a sobering report from the Intergovernmental Panel on Climate Change generated its own modest swirl of media attention.[4] *Vanity Fair*'s second annual "Green Issue" (May 2007) represents a glossier strand of climate change coverage. The cover depicts the actor Leonardo DiCaprio in an arctic setting that presumably faces imminent danger from global warming. Handsome and stoic, he seems ready to save the adorable polar bear cub at his feet through the sheer power of celebrity.
I raise these examples not only to indicate that cultural producers are contributing to an important conversation within the larger culture, but also

to raise some of the problems that we face when doing so. Some critical responses to *An Inconvenient Truth,* for instance, raise issues that we need to take seriously. Perceptions of inaccurate use of data, over-earnestness, or alarmism can irritate potential allies and alienate potential audiences. Or take that *Vanity Fair* cover. A friend refers to this kind of imagery as "green porn": sexy but superficial environmentalism that might help popularize issues like climate change and so lead to meaningful action, but might also merely offer a negligible hit of instant gratification. I worry a bit that those of us who are trying to address sustainability from within the visual arts are threading the same needle. If sustainability or climate change become art *trends du jour,* we risk providing a palliative to ourselves and to our audiences without contributing much to artistic production, nuanced debate, or lasting social change. To adopt the environmental philosopher Arne Naess' language: beware of shallow rather than deep, in art as well as ecology.[5]

Artists and institutions should also be clear about the implications of our intentions. In its online materials about *Weather Report,* the Boulder Museum of Contemporary Art expresses hope that the art in this exhibition will "inspire new awareness of, discussion about, and action on environmental issues, with new possibilities for envisioning a sustainable future." Variations of these moving and important goals have motivated generations of cultural producers to make complex works of art that are meant in part to serve a greater good.[6] Indeed, such instrumental aims seem almost an inherent part of art that addresses pressing issues like climate change. Still, they require a gentle caution. If handled bluntly by artists or by institutions, they could flatten work or leach out criticality, paradox, mess, play, and poetry. We need to be careful that worthy aims do not diminish the possibility for profound aesthetic experiences in whatever media, processes, or sites artists choose to activate.

So, the question bubbles up again in a slightly different form: given such caveats and the vastness of an issue like climate change, what can we do? Exhibitions like *Weather Report* contribute by providing resources to artists who are grappling with climate change—especially important since this work often takes forms that are not rewarded in the market. Such projects bring artists together with other artists and with thinkers in other disciplines. They disseminate art to wide audiences both locally and internationally (through programs, books, media coverage, word of mouth). In the case of *Weather Report,* the project may also realize the organizers' hopes of catalyzing new thinking and action within their Colorado community as well as in that dispersed national/international audience.

Beyond such immediate impact, I hope that this exhibition might also prompt a broader inquiry. One way of answering the question, "What can we do in response to climate change?" is to raise a meta-question: "What other questions do we need to be asking right now?" The exhibition offers a prompt. We can use it to think about questions that might be posed to cultural producers, museums, and other social institutions, scholars, policymakers, and the public in order to help generate new ways of thinking about art, sustainability, and the intersection of the two. What questions might help this new field deepen and develop? In this context I will only offer two, both of which stem from my perspective as a curator of contemporary art at a university art museum.

First, how can we develop more sustainable practices within cultural institutions? I have written about this elsewhere,[7] but the question bears repeating in relation to climate change. Institutions like art schools and museums consume vast resources. Especially if we are going to show art that addresses climate change or other topics related to sustainability, we should consider ourselves ethically bound to thoughtfully assess how we make use of resources (including money, time, and good will as well as materials). This is not simple or easy, but is certainly necessary.

Second, how can we connect the dots? How might we develop more flexible structures and resources that not only bring artists, scientists, activists, and other citizens together for short-term or small-group projects, but also create contexts within which ever more lasting, meaningful, and far-reaching interaction, exchange, and action might occur and be shared with others? Universities offer great potential. In addition to the infrastructural and intellectual resources that they offer, specific academic disciplines could generate fruitful alliances.[8] Other kinds of activity could be developed around existing nodes of independent cultural infrastructure—entities like the Experimental Station in Chicago or the New York-based Precipice Alliance that support small-scale, critical cultural experiments.[9] Foundations, donors, and policymakers might also be encouraged not only to support large-scale projects (like, say, an international sustainable art think tank...), but also to provide more direct funding to artists who work in ways that are not easily supported by the existing commercial or institutional mechanisms of the art world. Myriad other examples could surely be developed.

So, to return one last time to the initial question: What can cultural producers, institutions, and audiences do in response to climate change?

Weather Report offers responses from one curator, one museum, and a number of artists, and in this essay I sought to build on their work by offering a few additional questions and strategies. To return to Andrew Bird's lyrics, we might also try quieting down every once in a while. Like the female character in the song, we need occasional pauses to reflect on the ever-changing and sometimes inscrutable interrelationships in which we are all embroiled (again, micro and macro, part and whole). And then we need to get up from that perch at the edge of the bed and act on the insights derived from those critical, reflective pauses. We need to get out into the fray, into those imperfect and ongoing processes of trying to align personal, aesthetic, and institutional lives with sustainable values, into those complex weather systems. Don't worry too much about spilling the wine.

Stephanie Smith is Director of Collections and Exhibitions and Curator of Contemporary Art at the University of Chicago's Smart Museum of Art and an affiliate member of the university's Department of Visual Arts. Much of her recent writing and speaking has centered on the intersection of art and sustainability; she curated the exhibitions Beyond Green: Toward a Sustainable Art *(2005) and* Ecologies: Mark Dion, Peter Fend, Dan Peterman *(2000).*

Since Smith joined the Smart in 1999, its contemporary exhibition program has centered on thematic projects that play off the museum's university context. Prior to joining the Smart as Associate Curator, Smith held curatorial positions at the Rice University Art Gallery and Contemporary Arts Museum, Houston. She holds a masters degree in art history from Rice University.

Endnotes

Thanks to Tim Collins, Greg Sholette, and Mark Scala for their assistance as critical readers of this text.

1 See *Beyond Green: Toward A Sustainable Art*, Chicago: Independent Curators International and Smart Museum of Art, University of Chicago, 2005). Through 2009, a PDF is available for free download at smartmuseum.uchicago.edu.

2 For an overview of ways that some of these debates are addressed in environmental ethics, see Stephen M. Gardiner, "Ethics and Global Climate Change," *Ethics*, April 2004, 555–600.

3 As a few very recent examples of attention of the former, the art magazine *Frieze* addressed art and the environment in its summer 2007 issue; major multi-day international conferences in Sharjah, Chicago, and Paris held during April and May 2007, addressed related topics. As for the latter, the Chicago Humanities Festival, a massive annual city-wide series of events, is focusing for the first time on a social issue and will address climate change and related topics under the umbrella title "Climate of Concern" during fall 2007.

4 The IPCC report (its third) documents scientific consensus that climate change is indeed linked to human actions and likely to cause serious harm to our planet and its people—particularly and disproportionately in equatorial regions—although the severity of impact would vary depending on population and development trends over the coming century.

5 Naess formed his influential ideas about "deep ecology"—which address whole interconnected ecological systems that include but do not privilege humans, and advocate principles such as "an equal right to live and bloom"—partly in opposition to what he deemed a "shallow ecology" centered on the "fight against pollution and resource depletion. Central objective: the health and affluence of people in the developed countries." See Naess's 1973 essay "The Shallow and the Deep, Long-Range Ecology Movements," reprinted in *Deep Ecology for the 21st Century*, ed. George Sessions (Boston: Shambhala, 1995):151–155.

6 I am thinking here of a wide range of practices, including activist interventions within specific places and situations, experiments with innovative social and scientific ideas, and works—in any medium or format—that expand human faculties by opening us to new experiences and perspectives.

7 Smith, Stephanie "'Alas for the Dreams of a Dreamer': Art Museums and Sustainability" in *Land, Art: A Cultural Ecology Handbook* edited by Max Andrews, London: RSA, 2006, 190–195.

8 As one example, philosophers working in applied environmental ethics are seeking meaningful responses to systemic environmental problems and attending to the myriad ways that ethical choices are inflected by real-world parameters. The field moves ideas beyond the academy and into service to a wider conversation and broader field of action. This framing of the role of the public intellectual certainly resonates with the motivations and dilemmas of many cultural producers who care about sustainability. See, for instance, Andrew Light and Eric Katz, *Environmental Pragmatism* (London: Routledge, 1996).

9 I owe special thanks to Dan Peterman for many conversations about the power of art to provide a free zone for cultural experimentation. My ideas about independent cultural infrastructure are also rooted in my Chicago context with spaces like Mess Hall (www.messhall.org) and the Experimental Station (www.experimentalstation.org, founded by Peterman). Web sites like Free Soil's (www.free-soil.org) also provide important networked/dispersed nodes.

Coming of Age on a Crowded Planet

—Andrew C. Revkin

While studying in London some 30 years ago, I had a habit of stopping by a little booksellers' fair that convened every lunch hour in the financial district. Merchants wheeled out wooden carts and drew back canvas covers, and those of us eager for a bargain, or simply curious, dug in. Among the crumbling, leather-bound remains of someone's literary estate piled high on one of the carts, I found a small volume called *The Physical Geography of the Sea*, by Matthew Fontaine Maury. It was a sea captain's guide to the basics of oceanography and meteorology, published in 1859 by Sampson Low, Son, & Co. Maury was one of America's first oceanographers.

The book sat on my shelf, largely unread, until around 1988, when I first started writing at length about the theory that humans were warming the globe by adding to the atmosphere's natural heat-holding greenhouse effect. The main human climate influence was carbon dioxide, an unavoidable byproduct of burning fuels and forests. I scanned the book's yellowing pages and marvelous foldout ocean charts, looking for relevant sections. I homed in on a passage that was stunningly effective at describing the workings of the "spherical shell which surrounds our planet," as Maury put it. The book, nearly 150 years old now, still speaks powerfully of the importance of treating the air around us with respect.

The atmosphere, Maury wrote, "warms and cools by turns ... the earth and the living creatures that inhabit it. It draws up vapours from the sea and land, retains them dissolved in itself, or suspended in cisterns of clouds, and throws them down again as rain or dew when they are required ... It affords the gas which vivifies and warms our frames, and receives into itself that which has been polluted by use, and is thrown off as noxious .... It is only the girdling encircling air, that flows above and around all, that makes the whole world kin. The carbonic acid [carbon dioxide] with which today our breathing fills the air, tomorrow seeks its way round the world. The date trees that grow round the falls of the Nile will drink it in by their leaves ... and the palms and bananas of Japan will change it into flowers. The oxygen we are breathing was distilled for us ... by the magnolias of the Susquehanna, and the great trees that skirt the Orinoco and the Amazon .... The rain we see descending was thawed for us out of the icebergs which have watched the polar star for ages, and the lotus lilies have soaked up from the Nile, and exhaled as vapour, snows that rested on the summits of the Alps."

That language, worthy of Thoreau or Leopold, was worth the price of admission. But there was more: "Hence, to the right-minded mariner, and to him who studies the physical relations of earth, sea, and air, the atmosphere is something more than a shoreless ocean, at the bottom of which he creeps along," Maury continued. "It is an inexhaustible magazine, marvellously adapted for many benign and beneficent purposes. Upon the proper working of this machine depends the well being of every plant and animal that inhabits the earth; therefore the management of it, its movements, and the performance of its offices, cannot be left to chance."

I went on to quote him in my first book on global warming, back in 1992. Since then, the science supporting the idea that humans have a hand on the thermostat of the Earth has only built. We have arrived at a time when, voluntarily or involuntarily, humans are indeed "managing" the atmosphere. For the moment, even with all of the recent green ferment, many scientists I talk to regularly see scant signs we are prepared to take on that kind of role. We've been pumping about 27 billion tons of carbon dioxide into the atmosphere each year of late (a percent or two more each year, actually), and close to half of what goes up stays up for a century or more, accumulating like unpaid credit card debt. Most climate scientists are very confident that recent warming is mainly human-caused. Although the extent of warming and related changes in this century remains unclear—it could be manageable or a calamity—almost all experts seriously engaged on the question agree that unfettered emissions pose an unacceptable risk of producing dangerous, costly, largely irreversible climate and coastal shifts.

In the debate over how to respond to human-caused warming, there has been a heap of "woe is me, shame on you" discourse these last few years. We like to worry and to place blame elsewhere—on big companies (ExxonMobil has been a ripe target) or politicians (President Bush has become something of a poster child for denial to many climate campaigners). But that is oversimplified. Corporations and elected officials exist because people crave their products or prefer their platforms. Despite conspiracy theories, we are not coerced to buy bigger cars, televisions, or houses, to fly ever farther and more frequently for business or pleasure.

Indeed, rather than recoil in fear or anger from the idea that we have become a global-scale influence on this "pale blue dot," as Carl Sagan once described Earth, my sense is we need to embrace that reality, however discomfiting. Having grappled for a quarter century with global warming and the underlying trends shaping the phenomenon—explosive growth in populations and their energy appetites—I've come to the conclusion that the turbulent phase we are in is not so much lamentable as inevitable.

It is a bit like puberty in that sense. As its moment on earth has dawned, *Homo sapiens,* one of nature's great experiments, has necessarily had to have a selfish muscle-flexing, cow-tipping phase of exuberant adolescence. It is only natural for a youth to break things, burn things, kill things, to be mean and even nasty sometimes, if only to learn how that attitude can bite back. Mistakes are not only inevitable, they are vital. Progress depends on experimentation and failure, both in nature and in the subset of nature called human affairs.

It's also not surprising to me that it has taken a long time for the reality of our planet-scale power to sink in. Science for centuries drew us away from the idea that we were at the center of the universe, and now we are told that we are the dominant influence on our planet's biology and climate.

The real question is whether we, as a species, can "grow up" in time to avoid a long list of regrets. But doing this on a global scale will be harder than the transition individuals go through as they come of age. A youth or student can draw on accrued wisdom conveyed by parents or teachers or experience as a guide to how to save money for later needs, drive with caution even when angry, and negotiate instead of fight.

In considering how to shape personal, local, national, and global affairs so we can head toward 9 billion people with a reasonably thriving planet, history is no longer a useful guide. Our existing economics is built on the concept of endless growth, but that concept is running up against the reality of life on a finite orb with ocean fisheries that have declined 90 percent in half a century, tropical forests that are falling at the rate of a New York State each year, and an atmosphere that will later this century almost assuredly have at least twice as much of the main greenhouse gas, carbon dioxide, as it did before the Industrial Revolution kicked into high gear.

Our existing politics and diplomacy remain parochial even as the globalization of pollution, commerce, and shifting populations render borders increasingly obsolete. The atmosphere doesn't notice if a ton of carbon dioxide comes from a smokestack in Beijing or a taxicab in Boston. The burst of construction of coal-burning power plants under way in China, seen as the greatest threat to the global atmosphere and climate, is underway in large part to drive an economy fueled by manufacturing floods of products for consumers in the West.

So we have to invent not only the future, but also the new mindset needed to invent the future. It is a mindset with fewer boundaries, both geographic and temporal. Despite recent efforts to make current events like Hurricane Katrina an icon of global warming, the biggest risks from projected climate and ocean changes driven by the greenhouse buildup still mainly face our grandchildren and their grandchildren and on down the line. As the British economist Sir Nicholas Stern put it, "The sting is in the tail."

Risks from the climate extremes that are likely to be amplified in a warming world also mainly face people in places that have had the least influence, so far, on the greenhouse-gas buildup. Residents of semi-arid places near the equator, small low islands, and Arctic outposts where retreating sea ice is a highway and larder are starting to grumble about this inequity. At the same time, wealth and technology are insulating the world's established industrial powers from climate hazards.

© Andrew Revkin

The new mindset, experts tell me, will require those in the already-wealthy world to pay more heed to the aspirations and situations of those living in deep poverty, with lives still short and incomes measured in pennies per day. Climate matters. Wildlife matters. But human development matters, too. Progress on the environment cannot come without progress on the profound poverty that still grips billions of people, mainly far from those of us in the comfortable north. Even as we debate what to do about our fossil-fuel addiction, some two billion peasants and rural farmers around the world would love a kerosene stove. Instead, they still gather firewood or dried dung to cook daily meals on un-vented indoor fires. More than 1.6 million, mainly women and their stay-at-home kids, die young of avoidable respiratory illness each year as a result.

One image haunts me as I consider this path toward 9 billion (which is largely unavoidable even with growing awareness of the need for population control). While writing on rain forests, I was walking down a street in Manaus, the large city that is the human heart of the Brazilian Amazon. I rounded a corner and came upon a funeral parlor. I took this picture.

This is not a remarkable sight in many parts of the world today. As long as funeral parlors routinely carry coffins in all sizes, we must remember that climate change is a symptom, not the disease, and that for most people on this planet, other factors still matter more. It is not just the atmosphere "that makes the whole world kin," but also the shared miracle, and universal value, of human life.

I am confident we can make this transition, have this coming of age, and still have a reasonably functional planet and not too much social disruption. After a long day of reporting, I usually go to bed pessimistic. But for some reason, most mornings I wake up with a smile.

We have work to do. But we should relish the job, not fear it.

Andrew C. Revkin has covered the environment for The New York Times since 1995. His prize-winning books include The North Pole Was Here, *the first account of global and Arctic climate change for the whole family, and* The Burning Season, *on the life and death of Chico Mendes and his struggle to protect the Amazon rain forest. Revkin has won almost every major award in science journalism, not only for his articles, but also his photography and the 2005 Discovery-Times documentary "Arctic Rush." Much of his work can be found online here:* www.nytimes.com/revkin.

The Artists and Their Work

Kim Abeles's work chronicles historical and contemporary issues through sculpture and installation. Her work on the environment emphasizes air pollution, which was brought to national and international attention in 1985 through a series of works, The Smog Collector *series, in which Abeles found a way to collect particulate matter onto stenciled images. She has curated numerous traveling exhibitions relating to AIDS, the Gulf War, air pollution, and activist art. Her work has been exhibited throughout the United States, as well as in Canada, Belgium, the Netherlands, Spain, Vietnam, Thailand, and Czech Republic. Abeles represented the United States in Fotografie Biennale Rotterdam (1992) and at the Cultural Centre of Berchem, Antwerp (1993).* Kim Abeles: Encyclopedia Persona, A Fifteen-Year Survey *was sponsored and organized by the Santa Monica Museum of Art, touring the United States and South America. Abeles has been awarded grants from the Andy Warhol Foundation and Peter Norton Foundation and fellowships from J. Paul Getty Trust Fund for the Visual Arts, Pollock-Krasner Foundation and the California Arts Council.*

Kim Abeles

Presidential Commemorative Smog Plates are portraits of the US Presidents from McKinley to Bush created from particulate matter (smog) in the polluted air. The cut stencils on dinner plates were placed on the artist's rooftop for varying lengths of time depending on the extent of each president's violation or apathy toward the distressed environment. Upon removal of the stencil, the presidents' visages in smog are revealed. These images are accompanied by gold hand-lettered quotes from each respective president about the environment and business.

McKinley became the appropriate starting point because his presidency occurred when two automotive factories were running at full production in Detroit, thus marking a lifestyle and transportation mode that currently causes 70 percent of our air pollution. Also appropriate for the format of commemorative plates, McKinley was the first president to be marketed to the American public through merchandising of his image on buttons, posters, and souvenirs.

Kim Abeles
"Presidential Commemorative Smog Plate (George Bush in 40 Days of Smog)"
Smog, porcelain, gold
10.5" diameter
1992

Lillian Ball is an artist and environmental activist working in New York. Water imagery has been a constant subtext in her work, but recent subject matter focuses specifically on environmental concerns. The current body of work, Leap of Faith, *uses photography and video installation to address issues concerning an ongoing wetland preservation project on the North Fork of Long Island. She is currently developing interactive game installations such as* GO ECO, *which is based on the ancient Asian game, Go. Ball's serious game uses video projections to guide players through a wetland preservation process. Ball has received numerous awards including two New York State Foundation for the Arts awards, a John-Simon Guggenheim Foundation Fellowship and a National Endowment for the Arts grant. Recent installations were included in* Groundworks: Environmental Collaboration in Contemporary Art *at Carnegie Mellon University Gallery and* The Drop *at Exit Art in New York City, as well as solo shows at Art Sites, Riverhead and the Queens Museum of Art.*

Lillian Ball

When the Swedish Embassy invited Ball to Jukkasjarvi to carve ice, she had to think about it only briefly. As an environmental artist, the primary focus of her work has been on wetland preservation. Having followed climate change issues for several years, she realized this was a rare opportunity to observe the developing situation above the Arctic Circle firsthand. "Clearly, any artwork that I might do there could only be about this most urgent challenge of our time," she says. "On arrival, snowy Lapland, mother of pearl sunsets, the Aurora borealis, and all the beauty of the North greeted me. But after speaking with several Sami people about how the tree line is rising, how reindeer herds are starving for lack of lichen, and how below-zero weather never lasts as long as before, climate change was apparent."

66 Degrees, 32 minutes, 50 Years is a history of the rapidly changing Arctic ice cap that projects into the future, from 1990 to 2040. It is a story told in ice on ice, as the dwindling ice cap maps are animated and projected on a carved ice "arctic circle." Finding the appropriate maps involved extensive research and was more challenging due to projection models. A square metal pool will catch the melt water that in turn accelerates the disappearance of the ice, as is actually happening in the Arctic. Subsequent exhibitions of the work will show the animation video directly on the saved Torne River ice water.

Lillian Ball
"66 Degrees, 32 minutes,
50 Years"
Map morph animations projected
on ice (shown in photo) or ice
water (as currently exhibited)
2007

Born in Berhampore, India, and educated in part in the United States, Subhankar Banerjee left a career in science to pursue his art in the Arctic. Since 2001, he has photographed extensively in the American Arctic, focusing primarily on land, resource wars, and indigenous human-rights issues. He sees the Arctic supporting communities of our species and other species with which we share this planet. For his Arctic work, Banerjee received the Cultural Freedom Fellowship from the Lannan Foundation and the Greenleaf Artist Award from the United Nations Environment Programme. In 2007, a solo exhibition, Subhankar Banerjee: Resource Wars in the American Arctic *will be at the Hood Museum of Art at Dartmouth College in Hanover, New Hampshire. His work is featured in an international traveling group exhibition titled* Envisioning Change, *which opened in Oslo, Norway, and in a group exhibition at the Tucson Museum of Art.*

Subhankar Banerjee

Since 2000, Banerjee has focused all his photographic efforts on indigenous human rights and land conservation issues in the American Arctic. While he was initially motivated to document the so-called "last American frontier," after seven years of intense engagement, he thinks about the Arctic very differently. In his mind, the Arctic has become the most connected land on the planet, a mirror in which myriad cultures and issues are reflected with clarity. Migrations of hundreds of millions of birds link the Arctic—literally—to nearly every landmass on the planet.

However, the Arctic has also become a symbol of a disconnect writ vast, as global warming, toxic migration, and resource wars have made the region a tragic, constant reminder of the consequences of our carbon footprint. Banerjee's work approaches the Arctic not from aesthetic or intellectual considerations, but simply from the experience of "land as home." His photography attempts to address many of the compromises we have complacently accepted as a normal part of progress.

Subhankar Banerjee
"Gwich'in and the Caribou,"
from the series "Blood in my Meat"
"Charlie Swaney and Jimmy John;
Danny Gimmel and Rocky John;
Jimmy John; Danny Gimmel"
Pigmented inkjet print laminated
and mounted on sintra
Polyptych with 4 prints,
each 31" x 39"
2007
Courtesy of the artist

Iain Baxter& is one of Canada's leading multi-media artists and has been a pivotal figure in visual arts for more than forty-eight years. Recognized as an icon of conceptual art, he is among the most thought-provoking and pioneering contemporary Canadian and international artists. His highly regarded photography, conceptual installations, and projects have earned him the label "the Marshall McLuhan of the visual arts" for his ability to make us question, laugh, re-examine, and think. Baxter&'s works have been collected by the National Gallery of Canada, the Art Gallery of Ontario, the Museum of Modern Art in New York and the Gemeentemuseum in the Hague. His achievements have earned him numerous awards for his insightful and provocative works about our consumer cultural, social, ecological, and political conditions. In 2005, Baxter& received the Canada Council Molson Prize for his lifetime achievement in the arts as "one of society's great innovators." He is represented by the Corkin Gallery, Toronto, Ontario.

Iain Baxter&

For over forty-eight years, Iain Baxter& has been exhibiting art that focuses on and critiques environmental, ecological, and sustainability concerns; technology versus humans and nature; corporate, consumer, and popular culture; the rural versus the urban; and the concept of landscape in the broadest sense. *Animal Preserve #8* deals with the increasing concerns about the eradication of our animal species. Will they all become just preserved specimens in jars? *Killer Still-life #5* deals with the continuing use of deadly chemicals to kill every living entity. The residue of this practice is constantly affecting our water, weather, and environment. Still-life has played a major role in the history of art and *Killer Still-life #5* continues that tradition. But, in this particular work, the word "still" or "nature morte" takes on a very powerful meaning when these killing chemicals are used. It is imperative to understand how everything is connected in the web of life and to protect the animals and plants that share this planet, otherwise the net will dissolve.

Iain Baxter&
"Animal Preserve #8"
Mixed media installation
26" x 9" x 4"
1999-2007
Courtesy of the artist and
Corkin Gallery, Toronto,
Canada

Bobbe Besold is a working artist and activist who has exhibited and performed in the United States, Canada, China, Russia, Mexico, and England. Her work includes public art, photography, installations, sculpture, performance, film, painting, event design, set/costume/prop design, theater direction, writing, and education. Among numerous honors, she received the 2005 Grant Award from the John Anson Kittredge Educational Fund in Cambridge, Massachusetts; four full fellowships to the Vermont Studio Center in Johnson, Vermont; grants from The Puffin Foundation in Teaneck, New Jersey, and New Mexico Arts, Santa Fe Arts Commission; the Willard Van Dyke Award; and a full fellowship to the Bioneers Conference in San Rafael, California. As an independent curator and artist, Besold has overseen acclaimed exhibitions, performance events and installations, including Biophilia, Mutant Future, Animals of the Heart, Peace Day, Winterlight, The Summer Solstice, El Dia de los Muertos *(5 events)*, The International Pinhole Photography Exhibit and Station to Station *(Toronto metro)*.

Bobbe Besold

The Thirteenth Tipping Point is a Day of the Dead masked, costumed procession/performance event working in collaboration with the Boulder community. The performance uses humor, art, and ritual to illuminate our personal responsibility toward climate change. It examines death and transformation, denial and action. As Gary Snyder wrote in *Resurgence* magazine, a UK publication, "Performance is currency in the deep world's gift economy. The 'deep world' is of course the thousand-million year old world of rock, soil, water, air and all living beings..."

The project is in collaboration with Michelle Ellsworth and the Boulder Community.

Bobbe Besold
"The Thirteenth Tipping Point"
Performance
2007
Artist's rendering of
site-specific artwork created
for "Weather Report: Art and
Climate Change,"
Boulder, Colorado

Cape Farewell is a collaborative of artists, scientists, and educators who collectively address and raise awareness about climate change.

Created by David Buckland, Cape Farewell leads expeditions into the wild, beautiful, and icy High Arctic, a place for artistic inspiration and scientific inquiry.

Cape Farewell

On board the 100-year-old Dutch schooner, *The Noorderlicht,* Cape Farewell has sailed right to the heart of the climate change debate. From this vantage point the artists and scientists aim to illustrate the workings of this crucial part of the planet, drawing attention to the role ocean currents play and the effect rising CO_2 levels and changing weather patterns will have on us all and our climate.

Since 2002 Cape Farewell has led three voyages to the Arctic leading to a critically acclaimed exhibition, a film made in partnership with the BBC that screened on the Sundance Channel, a CD called *ARCTIC*, and its first book *Burning Ice*. Cape Farewell is quickly establishing itself as an international organization, with ambitious plans to engage with countries outside the United Kingdom, particularly the United States and Canada. In September 2007 Cape Farewell launches two voyages to the Arctic. International artists and scientists will explore anew the impacts of climate change on this and the wider global environment. A film documenting the voyages and innovative artworks developed by the artists will be shown internationally.

Cape Farewell
Documentation of
2001 expeditions

David Buckland/Cape Farewell
"Cold Library of Ice"
Color photograph
2001

Dan Harvey/Cape Farewell
"Carving Ice Credit"
Color photograph
2001

Mary Ellen Carroll is a conceptual artist who lives and works in New York City. Her art is infused with a wry sense of humor, moving deftly among large-scale architectural interventions, experimental films, staged events, sculpture, photography, and book projects. The work interrogates the relationship between subjectivity, language, and power. Consciously working without a signature style, Carroll creates work that is unified by its conceptual premises and its dedication to political and social critique. Carroll is the recipient of numerous grants and honors, including a Pollock/Krasner Award, a Guggenheim Fellowship, a Rockefeller Fellow and a MacDowell Colony Fellow. Her work has been exhibited at galleries and institutions around the world, including the Whitney Museum in New York City, the Institute of Contemporary Arts in London, Museum fur Volkerkunde in Munich, the Institute for Contemporary Art in Philadelphia, Vienna's MOMUK and the Renaissance Society in Chicago. Her work resides in numerous public and private collections.

Mary Ellen Carroll

Mary Ellen Carroll's installation *indestructible language* features a 900-foot-long neon sign that spells out the phrase, "IT IS GREEN THINKS NATURE EVEN IN THE DARK." It was originally installed across the five buildings of the American Can Company along the Pulaski Skyway in Jersey City, New Jersey, as the inaugural work for the Precipice Alliance. The collection of words frames the underlying moral problem and illuminates how we may choose to remain in the dark despite what we know about global climate change. The phrase's characters are created by 8-foot-high carbon neutral neon lighting elements manufactured by Lite-Brite Neon with low-wattage, energy-efficient transformers, and lead-free glass tubing all supplied by eco-materials pioneer Tecnolux and offset by donations from the Bonneville Environmental Foundation. Chester Jenkins of Village Type Foundry designed the energy-efficient font especially for *indestructible language*. The signage was illuminated only in the dark and was visible by airplanes flying into Newark International Airport and by cars driving onto the Pulaski Skyway from the New Jersey Turnpike.

Mary Ellen Carroll
"indestructible language"
Installation in Jersey City,
New Jersey
Neon signs
900' x 8'
2007

Center for Land Use Interpretation (CLUI) is a non-profit research organization that embraces a multi-disciplinary approach to understanding the nature and extent of human interaction with the earth's surface. Some of CLUI's programs include exhibits, tours, the Independent Interpreters *program,* Extrapolative Projects *in the field, residency programs, and publications. The Center houses a land use database that consists of a collection of source material and processed information on unusual and exemplary land use in the United States and is coupled with the CLUI Photographic Archive, a collection of thousands of images covering all types of land use sites. The* American Land Museum *is a network of landscape exhibition sites being developed across the United States to provide regional interpretive programming for the selected district it represents. CLUI was founded in 1994 and has produced over 30 exhibits on land use themes for public institutions all over the United States and overseas.*

Center for Land Use Interpretation (CLUI)

These images are from the Center for Land Use Interpretation's Photo Archive. The photos, depicting houses on the west end of Dauphin Island in Alabama, were taken November 2005, two months after Hurricane Katrina. The west end of Dauphin Island is a tenuous sand bar, a barrier island in the Gulf of Mexico, developed with numerous vacation homes. It is an area that is constantly battered by hurricanes, and where the land, such as it is, would have washed out from under these homes if not for artificial replenishment. These houses, many of which have been destroyed and rebuilt several times in recent history, seem barely connected to the place they occupy, and represent the outer limits of habitability in the nation's disintegrating margins.

Center for Land Use Interpretation (CLUI)
"Dauphin Island, Alabama, November, 2005"
Center for Land Use Interpretation Photo Archive
Inkjet print on paper
16" x 20"
2005

Brian Collier's integration of art and science explores the intricacies of a variety of environments by examining how human interaction transforms them. He works in a variety of media including sculpture, photography, video, performance, drawing, artist's books, and websites. In 2007, Collier's work is being exhibited at Neues Museum Weserberg Bremen in Germany, Chicago's Roland Contemporary and I-Space Gallery, and other locations. The College of St. Catherine (in St. Paul, Minnesota) and Washington University (in St. Louis, Missouri) held two of his 2006 shows. Collier was awarded a 2004 grant from the Illinois Arts Council for a solo show at Adams State College in Colorado. Collier received prior grants from the New York Foundation for the Arts (for a solo show in Havana, Cuba) and the City of Bloomington Cultural District, Illinois. He won CEPA Gallery's Emerging Artist Competition in Buffalo, New York, in 2001. A native of Long Island, he currently lives and works in Kansas City, Missouri.

Brian Collier

There is trouble on the horizon for the American pika. This small alpine animal may be one of the first species to go extinct due almost exclusively to the effects of climate change. Recent research in the Great Basin (the area between the Sierra Nevada and Rocky Mountains), has shown the disappearance of pikas from nearly 30 percent of sites that they were known to occupy within the last century. Pikas are particularly vulnerable to climate change because of their extreme sensitivity to temperature fluctuation. Temperature increase, changing patterns of vegetation, and lack of snow cover, which they rely on for insulation in the winter, may all be contributing to their decline. There is evidence that pikas have been moving ever higher to evade warming and drying in the region. As they reach the top of their mountain habitats, there will be nowhere else for the pikas to go, and we are likely to lose this amazing little creature.

The Pika Alarm is intended as both a warning and a call to action. Think of the pika as a canary in a coal mine, warning not only of the dangers it faces but of the broad effects that climate change is having on the global ecosystem. There is more evidence everyday that we are the primary cause for these dramatic changes, but we are also in the position to stop contributing to the problem and take some steps toward slowing down these devastating consequences.

The Pika Alarm is in collaboration with Chris Ray and Shana Weber.

Brian Collier
"The Pika Alarm"
Audio-rigged sculptures:
fiberglass, steel, wood, horn
speaker, motion-triggered audio
player, informational postcards,
website: pikalarm.net
2007
Site-specific artwork created
for "Weather Report: Art and
Climate Change,"
Boulder, Colorado

Xavier Cortada has worked with groups across the world to produce numerous large-scale collaborative art projects, including eco-art installations on Miami Beach and the South Pole, International AIDS Conference murals in Switzerland and South Africa, peace murals in Northern Ireland and Cyprus, and child welfare murals in Panama and Bolivia. The Miami artist has been commissioned to create art for the White House, the Florida Supreme Court, Miami City Hall, the Miami Art Museum, and the Museum of Florida History.

Cortada, a National Science Foundation Antarctic Artists and Writers Program grantee, created art for permanent display in McMurdo and Amundsen-Scott South Pole Station. Cortada's work is also in the permanent collection of The World Bank. He has received numerous awards for his volunteer work, including the prestigious Millennium International Volunteer Award from the US Department of State/ USA Today and the National Champion for Children Award from the Child Welfare League of America.

Xavier Cortada

On January 4, 2007, the 50th anniversary of the opening of the South Pole station, Cortada arrived at the South Pole. In his work, the Miami artist used the three-kilometer-thick glacial ice sheet that covers the South Pole as an instrument to mark time. At the geographic South Pole, Cortada planted an ice replica of a mangrove seedling on the ice sheet that moves approximately ten meters a year along the 60°W meridian.*

Embedded in the moving glacier, the "seedling" is sliding downhill in the direction of the Weddell Sea, 1,400 km away. The "seedling" has begun its 150,000-year journey towards the coastline, where it can eventually (theoretically) set its roots.

Through *150,000-year Journey*, the artist invites viewers to reflect on our role on this planet. Juxtaposing geological time frames with human time frames (e.g., the fifty-year history of human presence in the South Pole captured in *The Markers* project**), Cortada reaffirms the notion that we are merely custodians who should live in harmony with nature during the brief time we are here on earth.

*The "seedling" was a replica because foreign species cannot be introduced into the continent. As such, the artist created an ice sculpture using a mold made from an actual red mangrove seedling from Miami's Biscayne Bay.

**In *The Markers*, Cortada planted fifty-one different colored flags along 500 meters on the moving ice sheet that covers the South Pole. Each flag stood ten meters apart, marking where the geographic South Pole stood during each of the past fifty years (when humans first inhabited the South Pole). Each flag displayed the coordinates of the location on Earth where an important event took place during that year.

Xavier Cortada
"The Markers" (South Pole)
2007

Gayle Crites was born in Denver in 1949 and received her BA in printmaking from Colorado State University on a full-tuition Visual and Performing Arts Scholarship in 1971. Her career progressed to a regional level during the 1980s, when galleries in Santa Fe and Los Angeles first represented her paintings. She had her first national exhibition in New York in 1994. Crites currently lives and works in Boulder and exhibits in Chicago, Santa Barbara, Tucson, and Santa Fe. Although she painted in oil for many years, her most contemporary work focuses on large-format Japanese papers. Her multimedia work on paper is characterized by many layers of imagery in which she uses wax, brush, and ink drawings, thread embroidery and a variety of found-item collage elements. A recent trip to Japan impacted her on-going concerns about global warming, and many of her paper pieces consequently feature complex drawings about related issues.

Gayle Crites

During a trip to Japan during 2006, Crites observed evidence of a culture taking serious political and industrial positions on greenhouse emissions. Japan hosted the 1997 international conference, the Kyoto Protocol, which obliged thirty-five industrial nations to reduce their emissions of so-called greenhouse gases by 5 percent below 1990 levels by 2012. Crites created the drawing *Kyoto Protocol*, while she was in Kyoto, to depict a complex society infused with aesthetic and spiritual concerns. The drawing is about public awareness and pro-activity as well as the past, present, and future. *Biota Futurama—Recipe for Bio-Fuel* supports global scientific research to develop fuel alternatives from organic matter, such as algae. *Polar Bear Wreath* is a visual shrine to the retreat of winter sea ice caused by global warming, which has resulted in a decrease of the polar bear population.

Gayle Crites
"Kyoto Protocol"
Brush and ink on Japanese
handmade paper
Diptych, 16" x 44"
2006
Courtesy of 15th Street Studio,
Boulder, Colorado

Agnes Denes is known as a pioneer of conceptual and environmental art. She is best known for Wheatfield—A Confrontation, *a two-acre wheat field planted and harvested in Manhattan, and* Tree Mountain—A Living Time Capsule, *the largest reclamation site on earth, which reaches four centuries into the future. Her many awards include the Rome Prize (American Academy in Rome), The McDormott Award (MIT), The American Academy of the Arts and Letters Purchase Award, an Honorary Doctorate from Ripon College in Wisconsin, four National Endowment Fellowships, and the Watson Award (Carnegie Mellon). Her work is in collections of major museums including The Museum of Modern Art, Metropolitan Museum of Art, and the Whitney Museum of Art in New York; the National Museum of American Art, Corcoran Gallery of Art, and the Smithsonian Institution in Washington, D.C.; Philadelphia Museum of Art; Modern Museet in Stockholm; Kunsthalle in Nurnberg; Israel Museum in Jerusalem; and Wexner Center for the Arts at Ohio State University. Denes is a fellow at the Studio for Creative Inquiry, Carnegie Mellon University, Massachusetts Institute of Technology, New York University, and DAAD in Berlin.*

Agnes Denes

"A huge manmade mountain measuring 420 meters long, 270 meters wide, 28 meters high and elliptical in shape was planted with eleven thousand trees by eleven thousand people from all over the world at the Pinziö gravel pits near Ylöjärvi, Finland, as part of a massive earthwork and land reclamation project by environmental artist Agnes Denes. The project was officially announced by the Finnish government at the Earth Summit in Rio de Janeiro on Earth Environment Day, June 5, 1992, as Finland's contribution to help alleviate the world's ecological stress. Sponsored by the United Nations Environment Program and the Finnish Ministry of the Environment, *Tree Mountain* is protected land to be maintained for four centuries, eventually creating a virgin forest. The trees are planted in an intricate mathematical pattern derived from a combination of the golden section and the pineapple/sunflower system designed by the artist. Even though infinitely more complex, it is reminiscent of ancient earth patterns.

"*Tree Mountain* is the largest monument on earth that is international in scope, unparalleled in duration, and not dedicated to the human ego, but to benefit future generations with a meaningful legacy. People who planted the trees received certificates acknowledging them as custodians of the trees. The certificate is an inheritable document valid for twenty or more generations in the future. The project is innovative nationally and worldwide—the first such undertaking in human history. This is the very first time in Finland and among the first ones in the world when an artist restores environmental damage with ecological art planned for this and future generations. *(continued pg. 123)*

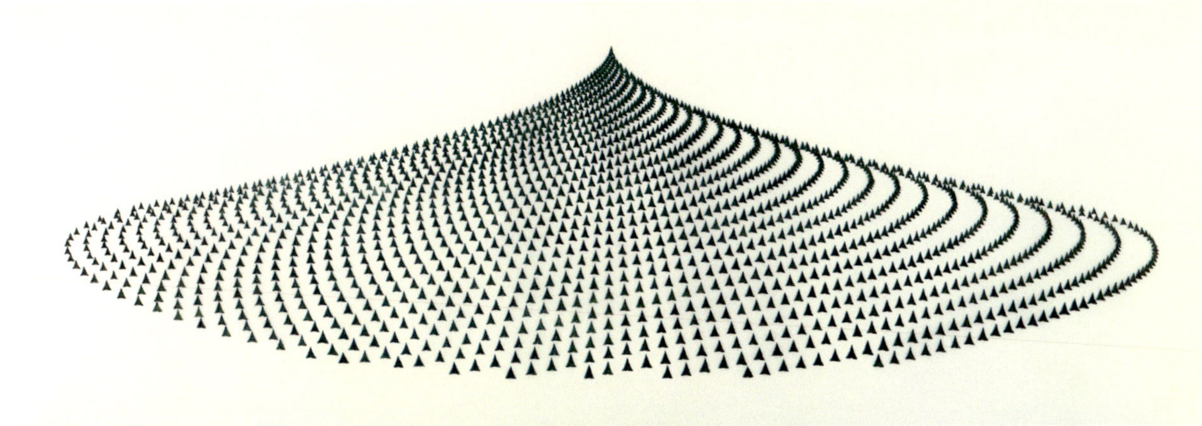

"Tree Mountain - A Living Time Capsule—11,000 Trees, 11,000 People, 400 Years," 1992-96, (420 x 270 x 28 meters), Ylöjärvi, Finland; Eleven thousand trees were planted in a complex mathematical pattern by eleven thousand people from around the world, to be maintained for 400 years. One of the largest reclamation sites in the world, *Tree Mountain,* created from refuse material of a mine, was declared a national monument to serve future generations with a meaningful legacy. Dedicated by the President of Finland, dignitaries and participants from around the world, 1996. *Tree Mountain* will become the first manmade virgin forest.

Top image: View of original design drawing from 1983, 35" x 96"
Bottom image: Winter view from 2001 of actual mountain/forest with the mathematical pattern beginning to show
Color photograph on foam-board
9" x 24"
1983, 2001

As a grandson of John Brown, a hereditary Euchee Chief, Steven Deo believes the continuum of Euchee culture is his responsibility and contribution. Deo's work is included in The Legacy Project, *a global exchange about the enduring consequences of many historical tragedies and in the PBS documentary,* Alcatraz Is Not An Island. *He has exhibited at numerous venues throughout the United States, including the Museum of Art and Design (New York), the National Museum of the American Indian (New York), the Smithsonian Institution (Washington, D.C.), the Orange County Center for Contemporary Art (Santa Ana, California), Gallery International (Baltimore), and Lois Lambert Gallery (Santa Monica, California). Deo received the Joan Mitchell Foundation Artist's Grant in 2005 and in 2003 received the Lorenzo de Medici award for sculpture at the Biennale Dell'Arte Contemporanea in Florence, Italy. Deo holds a Bachelor of Fine Art from the San Francisco Art Institute. He currently resides and works in Los Angeles County.*

Steven Deo

The artwork *bare trees* are images made from encyclopedia pulp (knowledge). The concept of the work depicts the transition of a tree to printable information (our quest for hard copy) and back to a tree (one without life). Although *bare trees* is unable to engage as part of our life support system, it functions as a reminder of trees' coexistence in our lives.

The four trees rendered in the installation are the *tree of knowledge*, the *tree of life*, a tree called *heartwood*, and the *tree of origin* as depicted in the Bible's Book of Genesis.

Steven Deo
From the series "bare trees"
"tree of knowledge"
Wood, encyclopedia pulp
Each 72" x 48" x 48"
Photo credit: Gene Ogami

Rebecca DiDomenico was born in Greenbrae, California. She attended school at California's Claremont College and Tribhuvan University in Nepal, then graduated from the University of Colorado with a Bachelor of Arts in English literature. "I am interested in unexpected relationships, the way a spider web mimics a wheel, the commonality between Dr. Seuss and the Dalai Lama, the resemblance between patterns in the constellations of the stars and the minute particles inside the human body," she says. Her work has appeared in many exhibitions, collections, and publications, including Boulder Museum of Contemporary Art, Museum of Contemporary Art in Denver, Denver International Airport, San Francisco Craft & Folk Art Museum, Denver Art Museum, Artspace in San Antonio, Art Papers, Artweek, Marin Independent Journal, *the* Washington Post, *and the* New York Times.

Rebecca DiDomenico

Shape-shifters have always been prevalent in the natural world. Nature is a giant testing ground to see which experiments fail and which succeed.

In the face of climate crisis, creatures will need to change drastically to survive. Life is a powerful force and while many species have succumbed to extinction and many more are endangered, it is not a new phenomenon for creatures to have to adapt to a variety of environmental changes. "I am not condoning standing back," DiDomenico says. "On the contrary, we have to act now to prevent destruction."

Shape-shifters both reflect ways animals have adapted and bring some futuristic options forward. Evolution has millions of ways to alter variables to create a different outcome. That is at the root of the true intersection of science and art—the faith and effort of experimentation. We need to be open to emergent ideas in times of great need. Ultimately, art and science have the ability to inspire hope, and therefore action, in others.

DiDomenico collaborated with Andrew Martin, an evolutionary biologist, who made suggestions as well as fine-tuned ideas.

Rebecca DiDomenico
"Intentional Mutation"
Transparency emulsions
mounted on Plexiglas,
glass marbles
11' x 42" x 2.5"
2007

Amy Franceschini is a San Francisco-based artist who focuses on the conflicts of over-consumption, pollution, and sustainability. Her work includes websites, installations, open-access laboratories, and educational formats that serve as platforms to discuss the politics of space and the effects of globalization. Franceschini founded Futurefarmers, an international collective of artists, in 1995, and co-founded Free Soil, a collective of artists, researchers, activists, and gardeners. Amy received her BFA from San Francisco State University and MFA from Stanford University.

Michael Swaine is dedicated to working in the community. One of Swaine's recent projects, titled Reap What You Sew Generosity Project, *consisted of the artist's pushing an old-fashioned ice cream style cart on wheels equipped with a treadle-operated sewing machine through the streets of San Francisco. On the 15th of every month he sews for free at the 509 Cultural Center at Cohen Alley. He received a BFA from New York State College of Ceramics at Alfred University and studied Advanced Ceramics and Sculpture at the Art Institute of Chicago.*

Futurefarmers

Amy Franceschini & Michael Swaine

The Unfinished Journey of Carl Linnaeus is a body of work anchored and inspired by the early expeditions of Carl Linnaeus and Alexander von Humboldt. Both taxonomists journeyed to the far reaches of the planet in search of new plant species. Humboldt included other artists on his journeys and Linnaeus depended on his own hand to tell the stories of what he saw with his eyes. Drawing upon the unique social situation of a scientist and artist working together to illustrate the natural world, *The Unfinished Journey of Carl Linnaeus* illuminates the social complexities of these relationships by inviting triads (three persons from very different perspectives) into small quarters, fashioned after the hull of a small research craft, to discuss the current state of the natural world. A stage built in Central Park, Boulder accommodates these triads. The frame of the fairy tale is told to each triad and they are asked to finish the tale. In the fairy tale, three people escape to a remote location to solve the world's problems. They drop out of the current carbon-dependent system to write their life's work and a "plan" to save humanity. Each group is invited to bring five items with them. Some may choose survival materials; others may choose monuments to the carbon culture they will leave behind. The first invited triad is made up of an environmental scientist, an urban planner/policymaker, and a gardener. In the remote location that they choose, the three will commiserate and document a sustainable plan for the future. Looking at such options as sustainable systems and open source policy, they have to survive not only philosophical differences, but also weather and the harsh elements that this location will present.

1

2

Futurefarmers (Amy Franceschini and Michael Swaine)
"The Unfinished Journey of Carl Linnaeus"
Performance, video, sculpture, drawing
2007
Site-specific artwork created for "Weather Report: Art and Climate Change," Boulder, Colorado

From "The Unfinished Journey of Carl Linnaeus"
"Starting Points: 1. Inside Research Vessel and 2. Researchers' Backpacks"
Collage/Drawing
11" x 14"
2007

Bill Gilbert is co-director of Land Arts of the American West at the University of New Mexico (UNM). Land Arts is an interdisciplinary field program expanding the definition of land art through direct experience with the full range of human interventions in the landscape—from pre-contact indigenous to contemporary practices. Gilbert will work with UNM biologist Bruce Milne and University of Colorado at Boulder hydrologist Vijay Gupta to examine plant and animal communities in the interface of mountain and prairie ecosystems in terms of the hypothetical changes expected as a result of global warming.

Bill Gilbert

Bough (bou), n. –A fractal-based planting design for the mountain/prairie ecotone above Boulder, Colorado. The design utilizes plant species capable of adapting to the rising temperatures brought on by global warming.

Location: N39° 58′45, W105° 16′22
Elevation: 6074′
Plant Total: 516
Species: Gambel Oak, Mountain Mahogany, Piñon Pine

The project is in collaboration with artist Erika Osborne and biologist Bruce Milne.

Bill Gilbert
"Bough"
Digital print on Plexiglas
40" x 40"
2007
Proposed site-specific artwork
created for "Weather Report:
Art and Climate Change,"
Boulder, Colorado

Isabella Gonzales is a native of the border city of El Paso, Texas. This cross-cultural setting is often reflected in her work, as are issues affecting immigrants and people living below the poverty level. She has worked as an artist and an educator for most of her adult life and has been active in numerous causes over the years, including domestic violence, human rights, the environment, and the peace movement. Gonzales shows her work regionally and was included in the New Mexico chapter of the National Museum of Women in the Arts at the Museum of Fine Arts in Santa Fe in 2005. She also completed a site-specific installation commissioned by the Rubin Center at the University of Texas at El Paso in 2006. In addition to maintaining a studio in Galisteo, New Mexico, she works for the Adelante Program that serves homeless children in the Santa Fe public schools.

Isabella Gonzales

The ozone layer is roughly the thickness of two pennies. This piece invites everyone to add their two cents' worth by dropping two pennies into a penny jar. Viewers are also asked to take action by filling out a postcard addressed to legislators and by taking postcards that contain information on environmentally friendly everyday choices and mailing them to friends and family. Kitchen table discussions are frequently the setting for the beginnings of big changes. Our children often bring up questions we are too tired to consider. Presenting a penny jar as a symbol for change highlights their future, which is very much dependent on our action—or inaction—today.

Isabella Gonzales
"2¢ worth...your participation is requested"
Mixed media installation
Variable
2007

Over the last five years, the University of Colorado's (CU) College of Architecture and Planning has been one of the leaders in green building and design. CU won the first two Solar Decathlon competitions, took four of the five top places in the Colorado Chapter level of the United States Green Building Council's (USGBC) Natural Talent Design Competition in 2006, and was nationally recognized with a Honorable Mention at Greenbuild, the USGBC's Annual Green Building Conference in November 2006. Locally, CU has been instrumental in the advancement and education of green design, winning the CU Environmental Center's design competition for a solar addition to the Boulder Campus Recreation Center and taking two of the top three awards in the "Barkitecture 2005," a green doghouse design competition benefiting the Boulder Humane Society. With a committed faculty, CU's College of Architecture will continue to look for solutions to reduce building impact on the environment.

Green Fabrication

University of Colorado, College of Architecture and Planning

The aim of this work is to explore furniture design in a more symbiotic relationship with the natural environment. Assessing this relationship begins with the research of the material's origins, which is then constantly evaluated until the furniture's ultimate deconstruction. Students develop a response toward sustainable principles and construct a prototype that explores green materials, zero-waste production, closed-loop recycling, and furniture designed as biological or technical "nutrients."

This work represents a starting point for more sustainable design but does not attempt to proclaim a solution. Inevitably, each student, forced to compromise at some point in the construction process, makes difficult decisions about how his or her piece will ultimately impact the environment. The larger lesson of this process is how seriously our choices impact the planet.

Green Fabrication
(courtesy of Erik Sommerfeld,
Professor of Green
Fabrications, University
of Colorado, College of
Architecture and Planning)

Student work:
Bill Daher
"Square Chair"
Poplar wood veneer, low-VOC
paint, natural cork, post-
consumer waste aluminum
Prototype
2006

Megan Haughey
"Resting Place"
3form ecoresin
Prototype
2006

Kellen Schauermann
"Schauermann Worktop"
Powder-coated recycled steel,
bamboo plywood
Prototype
2006

The Harrisons, Helen Mayer Harrison and Newton Harrison, are Professors Emeriti from the University of California at San Diego. They have been collaborating as environmental artists since the early 1970s and were active in America then, but since the mid-1980s they have done a large part of their work in Europe. As principals of the Harrison Studio, they currently have a grant from the United Kingdom to create an exhibition on global warming and ways to deal with its possible effects. They will be returning to Terre Haute, Indiana, to work with a project that they proposed for the area in the early 1990s. They have exhibited works in the Venice Biennale, the Sao Paulo Biennale, Documenta 8, and museums in France, Germany, Holland, Croatia, Japan, and the United Kingdom. They have been associated with Ronald Feldman Fine Arts in New York City since 1974.

Helen Mayer Harrison & Newton Harrison

Some years ago, the artists were working in Europe and visited an old friend and coworker, Professor Dr. Georg Grabherr, Director of the Department of Vegetation, Ecology, and Conservation Biology at the University of Vienna. He showed them his latest project, presented as a series of informative posters of charts and texts with several small images. This project was the study of the effects of global warming on the vegetation of a nearby alpine mountain called Schrankogel. Each year, as the temperature slowly rises, several species of grasses and flowering plants that he and his postgraduate students were studying moved their habitats slowly up the mountain. The plants were slowly—and not so slowly—disappearing from the lower altitudes of the mountain and reappearing at higher levels as the temperature rose, until a few of them reached the top and disappeared. The Harrisons were intrigued and asked Dr. Grabherr if they could translate his study into a story.

Dr. Grabherr, as a scientist, was commissioned by the United Nations to repeat this study on a dozen or so mountains in other parts of the world.

<u>This is a little drama</u> entitled The Mountain in the Greenhouse.
The theme is the disruption living systems will undergo
as the perturbations of global warming
reverberate through the European high grounds.
It is a drama being enacted
in fast time if you happen to be a glacier
but in slow time if you happen to be a person.

Helen Mayer Harrison and
Newton Harrison
"The Mountain in the
Greenhouse"
Video
2001

Judit Hersko is an installation artist whose work has been featured in numerous exhibitions in the United States and Europe. In 1991 she was a visiting artist at the Women's Museum in Bonn, Germany, and in 1997 she represented her native Hungary at the Venice Biennale. Her work has been shown in Germany, Austria, Hungary, Spain, and in many cities around the United States, including Chicago, New York, Los Angeles, and San Diego. In 1995 she received an Artslink Collaborative Grant and in 1998/1999 she won a California Arts Council Visual Arts Fellowship in Sculpture. In 2005, she was invited as a fellow to the prestigious Lucas Artists Residency Program at Montalvo. She has several pieces in museum collections, including The Museum of Contemporary Art, Ludwig Museum in Budapest. Her work has been the subject of many publications including articles in Sculpture Magazine *and* Art in America. *She is currently an assistant professor in the Visual and Performing Arts Department at California State University, San Marcos.*

Judit Hersko

Seven Days of Dissolution is part of a larger installation titled *Shifting Baselines* that reflects on climate change in the context of related social and cultural phenomena.

The installation also depicts the research of biological oceanographer Victoria Fabry, who studies the effects of ocean acidification caused by the ocean's absorption of atmospheric CO_2. Dr. Fabry and her colleagues have found that calcifying organisms, including pteropods, corals, and shellfish, show reduced calcification in response to ocean acidification.

If CO_2 emissions continue to rise at their current rate, some ocean regions are expected to encounter surface seawater that is corrosive to these organisms by the year 2050. By 2100, the shells of many forms of marine life may dissolve outright in such regions. It is likely that biodiversity will be adversely affected unless the ocean acidification process is slowed and reversed.

Judit Hersko
"Seven Days of Dissolution"
Seven acrylic tanks, seven black sea fans with Capiz shell sculptures of human heart and lung dissolving, one carbon dioxide tank
Acrylic tanks each 12" x 12"
2006
Courtesy of the artist
Photo credit: Robert Steinberger

Colorado artist Lynne Hull has pioneered "trans-species" art, creating sculpture installations as wildlife habitat enhancement. Hull has worked in the American West with a variety of wildlife agencies including state wildlife departments, the Forest Service, the Bureau of Land Management and the National Park Service. She has worked in fourteen states and eight countries, with a wide variety of wildlife agencies and communities. In 1998 she worked in Yucatan and Chiapas, Mexico, with the environmental non-governmental organization Pronatura. In 1993–94 she realized three projects in the United Kingdom, courtesy of a special Fulbright Fellowship and one in Kenya on a Lila Wallace/Reader's Digest Foundation/Arts International Residency. Currently she is working on Migration Mileposts, *linking communities in the hemisphere that share migratory birds, and she recently completed* East Drake Pondworks, *a major Art in Public Places piece for the city of Fort Collins, Colorado, where she lives.*

Lynne Hull

Lynne Hull's research on biodiversity has made her aware of the impact that human-induced climate change will have on wildlife. Climate change will be major—threatening or endangering many species. Some animals, like bats and pikas, are already feeling the effects. Nursing mother bats must have immediate access to water to drink as they emerge, dehydrated, from their daytime roosts. Pikas are losing their food sources as high elevation grasses change due to warming. Issues of wildlife connectivity will be even more urgent as animals have to deal with changes in what food sources grow where, shifts in accessible water, and impacts of relocations of people.

To express these concerns, Hull has created *Wildlife Warning* signs and located them at points where open space, wildlife-friendly land meets the urban interface. Each would be a warning to wildlife about the danger they will encounter by entering the urban area, and a reference to climate change, which threatens water supplies for wildlife. Signs will also remind human observers about this challenge. Each sign is square and yellow and oriented at 45 degrees, like a street "warning" sign. On the center of each sign resides an "erased" local wildlife species. The image is drawn and then erased, mimicking the creation, evolution, and extinction of species. A water capture basin accompanies some signs.

Hull is also creating green "Cooler Choice" stickers where anyone seeing the reminder has a small choice to make which can impact climate change. The stickers indicate the choice that helps reduce global warming—using stairs instead of elevators, driving a fuel-efficient car, taking the bus, riding a bike, changing lightbulbs, and many more.

Lynne Hull
"Wildlife Warning: Global Warming"
Mixed media
18" x 18"
2007
Site-specific artwork created for "Weather Report: Art and Climate Change,"
Boulder, Colorado

Pierre Huyghe was born in Paris in 1962. He has had numerous solo exhibitions at international venues including Tate Modern, London (2006); Musée d'art Moderne de la Ville de Paris (2006); Moderna Museet, Stockholm (2005); Castello di Rivoli, Torino (2004); Dia Art Foundation, New York City (2003); Solomon R. Guggenheim Museum, New York City (2003); Kunsthaus Bregenz, Austria (2002); Neue Nationalgalerie, Berlin (2002); Stedelijk Van Abbemuseum, Amsterdam (2001); and Museum of Contemporary Art, Chicago (2000). His work has been exhibited in Documenta 11, Kassel (2002); 1999 Istanbul Biennial, Carnegie International, Pittsburgh (1999); Manifesta 2, Luxembourg (1998); 2nd Johannesburg Biennial (1997); and the Biennale d'Art Contemporain de Lyon (1995, 2003, 2005). In 2001 Huyghe represented France at the Venice Biennale, and in 2002 he received the Hugo Boss Prize from the Solomon R. Guggenheim Museum. He is represented by the Marian Goodman Gallery, New York City/Paris. He lives and works in New York City and Paris.

Pierre Huyghe

A Journey That Wasn't (2005) is inspired by tales of uncharted islands emerging from the melting ices of the Antarctic shelf. In February 2005, Pierre Huyghe and a handful of artists embarked on an actual expedition to Antarctica in search of a unique solitary white creature—an albino penguin, rumored to inhabit an unnamed island somewhere at the height of the Polar Antarctic Circle. Once discovered, the island served as a base for experimentations meant to facilitate contact with the animal. A machine set on the icy shore translated the island's topography into a musical score and it was the artist's hope that it would be a beacon for the encounter with the elusive creature. Once the expedition returned to land, the newly discovered island was staged and filmed at the Wollman Rink in New York's Central Park in October 2005, and the soundtrack created in Antarctica served as the centerpiece for a musical. The footage from this recreation—a negative version of the Antarctic landscape, with black ice, an albino penguin, and a forty-piece orchestra—is united with documentation from the original journey. The resulting single-film projection creates a new territory of reality where fact, fiction, and representation become one.

Pierre Huyghe
"A Journey That Wasn't"
Super 16mm film and HD
video transferred to HD video,
color, sound
21 minutes and 41 seconds
2005
Courtesy of the artist and
Marian Goodman Gallery

Basia Irland creates international water projects that are featured in Water Library *(University of New Mexico Press, November 2007). Irland, a University of New Mexico professor and founder of "Arts and Ecology" in the College of Fine Arts, often works with scientists from diverse disciplines creating rain-water harvesting systems; community projects along lengths of rivers; and waterborne disease projects around the world, most recently in Ethiopia, Egypt, and India. Irland is the recipient of over forty grants including a Senior Fulbright Research Award for Southeast Asia, Woodrow Wilson Foundation Fellowship Grant and a National Oceanic and Atmospheric Research Grant. She has produced five video documentaries, had twenty-seven one-person international exhibitions, and participated in hundreds of group shows. Essays about her work have been included in books published in Germany, Switzerland, and the United States. Her art is in collections around the world. She relishes hiking along streams, canoeing, and thanking collaborators.*

Basia Irland

A Gathering of Waters: Boulder Creek, Continental Divide to Confluence, is a collaborative project connecting hundreds of people along Boulder Creek's forty-seven miles (an action that can be replicated anywhere) and documents through video the necessity of communal effort and scientific knowledge to deal with the complex issues of climate change, including less snowpack, early snowmelt, and diminishing glaciers. Water gathered by participants from Boulder Creek is added to the canteen, decanted, and handed to others who collect more creek water in the canteen, illuminating the point that we all live downstream. This gathered creek water is then frozen, carved into the form of a 250-pound open book, and placed back into the creek. The book is embedded with rows of an "ecological language" or riparian "text" consisting of columbine, blue spruce, and mountain maple seeds, which are released as the ice melts into the current. When the plants grow along the bank they will help sequester carbon.

The scientist with whom the artist worked is biogeochemist Dr. Jason Neff. Collaborators include John Davis, Jessica Damon, Curry Rosato, Rick Silva, Talice Lee, and hundreds of Colorado participants, including Fergus Stone who wrote the commissioned song, "Ballad of Boulder Creek."

Basia Irland
From "The Gathering of Waters: Boulder Creek, Continental Divide to Confluence"
2007
Site-specific artwork created for "Weather Report: Art and Climate Change," Boulder, Colorado

"Boulder Creek Backpack"
47 glass vials (one for each mile of the creek), recycled car inner tubes, aspen branch, and sticks cut by beaver. The backpack repository contains objects from the four-month project including the logbook, canteen, video documentary, water samples, hydrology reports, watershed maps, and a USGS hydrograph for stream-flow in Boulder Creek at North 75th St., April 2005 to January 2007.
Photomontage of backpack being worn
23.5" x 50" x 3"
2007

"receding/reseeding"
250-pounds of ice, frozen from water gathered by project participants; small stones; columbine, blue spruce and mountain maple seeds.
Photographic documentation
20" x 40" x 10"
2007

For forty years, Patricia Johanson's multidisciplinary designs have combined art, ecology, landscaping, and functional infrastructure. During the 1960s and 1970s, Johanson created 150 gardens for House & Garden Magazine *and plans for Yale University, the town of Columbus, Indiana, and Con Edison. Her projects include* Fair Park Lagoon, *Dallas (1981–86), one of the earliest ecological artworks;* Endangered Garden *(1988–1995), which incorporates tidal sculpture, butterfly meadow, and habitat restorations into a baywalk/sewer along San Francisco Bay;* Amazon Rainforest Park, *Obidos, Brazil (1992) and* Nairobi River Park, *Kenya (1995), which feature sculptures that filter polluted water;* Ulsan Dragon Park, *Korea (1996) and* Rocky Marciano Trail, *Brockton, Massachusetts (1997), which combine ecological and sculptural playgrounds with flood control and watershed restoration; and* Millenium Park *in Seoul (1999), which transformed the world's largest landfill into a terraced guardian figure. Johanson's recent works include an urban garden for the French government, the 272-acre* Petaluma Wetlands Park *in California, and* The Draw at Sugar House, *a Salt Lake City highway crossing.*

Patricia Johanson

Structures created by wildlife, living organisms, and natural landforms have inspired Johanson's building designs since the 1960s. An article published in *The Princeton Journal,* "Architecture as Landscape" (1985), suggests abandoning the concept of buildings as "objects" in favor of "buildings that grow and change." Today the need to reduce energy and resource consumption demands design on an organic model.

Buildings that form alliances with living systems—earth, plants, water, wind, and sun—reap the benefits of a reconnected world where ecological communities, agriculture, and development are not antithetical to one another. The ebb and flow of breathing, seasons, internal heat, and natural cleansing subsume the hubris of ego and waste.

Patricia Johanson
"Living Apartment Houses"
Pencil, colored pencil
8.5" x 11"
1969

Chris Jordan is a Seattle-based photographic artist whose work examines the phenomenon of American consumerism. His work has been exhibited widely in the United States and Europe and has appeared in print media worldwide.

Chris Jordan

This series looks at contemporary American culture through the austere lens of statistics. Each image portrays a specific quantity of something: 15 million sheets of office paper (five minutes of paper use), 11 thousand jet trails (eight hours of US flights), and so on. The hope is for images representing these quantities to have a different effect than the raw numbers alone, such as we find daily in articles and books. Statistics can feel abstract and anesthetizing, making it difficult to connect with and make meaning of the real-life phenomena that they represent. This project visually examines these vast and bizarre measures of our society in large, intricately detailed prints assembled from thousands of smaller photographs.

"Office Paper" detail

"Office Paper"
Depicts 30,000 reams of office paper, or 15 million sheets, equal to the amount of office paper used in the US every five minutes.
Archival inkjet prints
60" x 87"
2007

In her sculpture, Marguerite Kahrl approaches systems thinking from a playful angle, manipulating messages and media to instigate social engagement. Her idiosyncratic constructions highlight our complex relationship with energy and the land. For example, Kahrl's remote-controlled vehicles describe a cycle of consumption. These elaborate machines are constructed from junkyard parts and operate on a by-product of organic waste: methane, a landfill gas. At first glance, her Meek and Timid Action Figures *resemble toys, yet the figures bear expressions of grief and panic, overwhelmed with the responsibility of caring for the bite-sized nuclear waste barrels, included in the accompanying Nuclear Cooling Tower Accessory Kit. Kahrl's role as a social sculptor can also be seen in her long-term experiment of an off-the-grid environment located in Maine. Kahrl received her Bachelor of Arts/Bachelor of Fine Arts degree from the University of Colorado at Boulder and a Master of Fine Arts degree from the Rhode Island School of Design. She has exhibited widely in Europe and the United States. She currently lives on a moraine in Northern Italy.*

Marguerite Kahrl

The installation *Homeland Insecurity* places artificial industrial hemp plants in planters at the entrance of the museum and encompasses them in a soundtrack of voices reading scientific research about industrial hemp. The soundtrack, which refers to the contested political history of industrial hemp in the United States, runs through an electronic filter that transforms the hemp research into a covert, subliminal message—a form of manipulation, concealed from the conscious mind that is often used as a tool in advertising and political campaigns.

In 1937, under pressure from the Hearst Paper Division and Dupont, Congress passed a prohibitive tax on the cultivation of one of the nation's premier renewable resources. Fearing competition to their timber and petro-synthetic products, the companies sought to cast industrial hemp as identical to marijuana. In contrast to the smear campaign of that era, *Homeland Insecurity* incorporates factual data to counter the soiled reputation of a crop with as-yet-underdeveloped environmental and economic benefits. Similar to dystopic satire or science fiction, the aim is to blend fact and fiction, causing us to question our present reality.

Marguerite Kahrl
"Homeland Insecurity"
Artificial hemp plants, planters, subliminal sound installation
242" x 88" x 90"
2007
Site-specific artwork created for "Weather Report: Art and Climate Change,"
Boulder, Colorado

"Study for Homeland Insecurity (a)"
Giclee print
11" x 17"
2007
Courtesy of the artist

Janet Koenig and Greg Sholette frequently collaborate, combining their graphic design and sculpture skills. Each project is extensively researched and planned for its unique location. Together they have created such works as Surviving Paradise *for Wave Hill in the Bronx (2006);* disLOCATIONS *for inSITE94 in San Diego, California, and Tijuana, Mexico; and* Installation *at the Alternative Museum in New York City (1986). Through the groups REPOhistory (1989–2000), and Political Art Documentation/ Distribution (PAD/D 1980–1988)—two public art and activist collectives—they were involved in a range of art projects in New York City, Atlanta, and elsewhere. Janet Koenig's solo project* Eat ... Shop ... Vote ... *was exhibited at the Ramapo College Library, Ramapo, New Jersey, in 1998, and* Greg Sholette Selected Projects *was exhibited with a catalogue at Colgate University, Hamilton, New York, where he was their Distinguished Batza Family Chair in Art and Art History in 2004. Sholette's most recent book,* Collectivism After Modernism: Art and the Social Imagination after 1945, *is co-edited with Blake Stimson and published by the University of Minnesota Press, 2007.*

Janet Koenig & Greg Sholette

An estimated 750 million PCs are in use this minute across the world. These PCs get discarded and replaced about every five years. The manufacture of just one average desktop PC and monitor requires "1.8 tons of raw material," according to a 2004 United Nation's study. The study also reports that when used PCs are not properly recycled, additional carbon emissions from discarded plastics and hazardous chemicals further contribute to global warming and illness. The aim of *Cannibal Tech* is to make visible the connection between the three elements involved in the cycle of PC production and waste: the technology industries whose production methods contaminate the earth's ecosystems, the global labor force whose health is compromised by the production and recycling of computer components, and the consumers (including artists and intellectuals) who buy and discard those machines—and who often overlook the way their "immaterial" labor contributes to the first two processes.

Cannibal Tech was made possible by The Puffin Foundation and with special thanks to Dan Matsch and Eco-Cycle Incorporated.

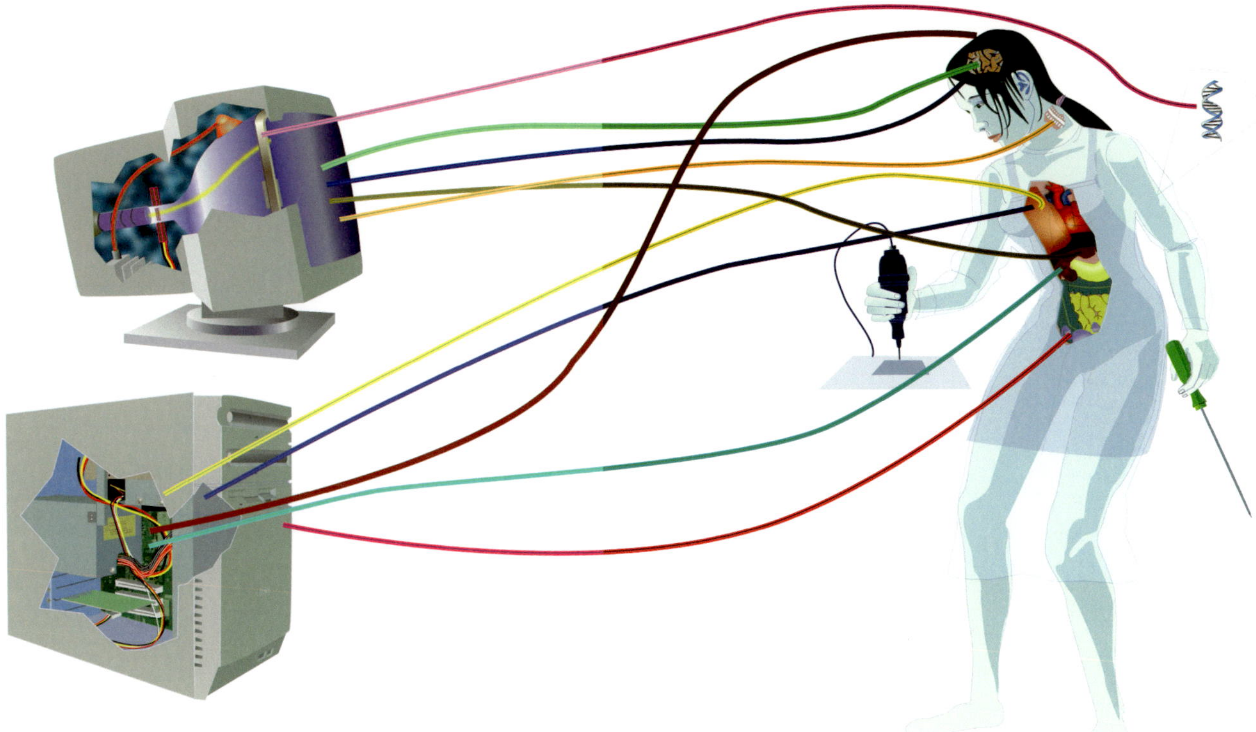

Janet Koenig and
Greg Sholette
"Cannibal Tech"
Inkjet poster, recyclable
computer parts
Each 4' x 4' x 8'
2007
Site-specific artwork created
for "Weather Report: Art and
Climate Change,"
Boulder, Colorado

Eve Andrée Laramée has been exploring the relationship between art, science and nature for over twenty years through sculpture, installation, and works on paper. Her work has been exhibited throughout Europe and the United States and is included in the collections of the MacArthur Foundation, the Museum of Modern Art in New York, the Museum of Contemporary Art in Chicago, the Fogg Art Museum of Harvard University, the Hammer Museum at the University of California, Los Angeles, and the Albuquerque Museum of Art. She has taught sculpture and critical thinking at Cooper Union, Massachusetts Institute of Technology, Rhode Island School of Design, Sarah Lawrence College, New York University, and Maryland Institute College of Art. Laramée has received awards from the Pollock-Krasner Foundation, the New York Foundation for the Arts, the Mid-Atlantic Arts Foundation, the National Endowment for the Arts, the Guggenheim Museum, the Andy Warhol Foundation, and MacDowell Colony. Laramée is the Chair of Interdisciplinary Sculpture at the Maryland Institute College of Art.

Eve Andrée Laramée

Pretty Vacant, a public artwork by Eve Andrée Laramée, addresses the environmental issue of desertification—the global degradation of land occurring in over 100 countries on all the populated continents. Desertification is caused by global and regional climate change and human behaviors, which contribute to climate change. Laramée's ephemeral public work is circulated through an existing system of the local mass media, the Boulder newspaper the *Daily Camera*, which reaches an audience of almost 100,000. The title of the piece refers to widespread cultural attitudes towards the desert as being "vacant" space, as well as to the apathy and rage of the 1977 Sex Pistols' song of the same title. Laramée's subversive deployment of the mass media as a distribution system, combined with a cut-and-paste visual sensibility (referring to her past employment in the 80's as a "paste-up" artist at *Artforum* and *Interview* magazines), combines her own photographs of desert lands with NASA satellite imagery and text-based information.

This "public service announcement" was made in collaboration with environmental scientist Carol Wessmann and designer Katherine Tsui.

Eve Andrée Laramée
"Pretty Vacant"
4-color ink on newspaper stock
Each 11.625" x 10.75"
2007
Site-specific artwork created for "Weather Report: Art and Climate Change,"
Boulder, Colorado

Learning Site comprises Rikke Luther (Denmark), Cecilia Wendt (Sweden), Julio Castro (Mexico) and Brett Bloom (United States). The group focuses on the local conditions it finds in the place where it chooses to work. In past years, Learning Site has mainly worked with resource materials and economies related to the specific situations where work has been carried out. They investigate economic, environmental, labor, property rights, and many other issues in tandem to produce a variety of perspectives. Learning Site takes part in a discussion of how knowledge is distributed and produced.

Learning Site

The poster describes the construction of a solar cell greenhouse called *Extension Module #001* constructed in Copenhagen Harbour. The *Extension Module #001* expands on an already existing floating dwelling and island. *The Extension Module #001* is an octahedron, composed of eight equilateral triangles made of wood, polycarbonate, and solar panels. The roof is designed to collect rainwater. The work includes a historical reflection of the solar cell, a discussion of the politics of solar energy (like economics, ownership, and regulation of the energy produced by the sun), and a brief comment on climate change predictions by the Swedish chemist Svante Arrhenius, who discussed the effects of human CO_2 emissions in 1896.

This project is a collaboration between chemist Maiken Vibe Bauer and Learning Site.

Where is the Sun?

- Gleaning the Solar Power

Through the increasing influence of CO2 in the air, we hope that we will gradually approach a time of better, more stable climates, especially in the colder parts of the world. This would make the earth able to yield larger harvests for the benefit for the quickly growing populations.

- Svante Arrhenius, 1906

[Extension Module #001], a floating greenhouse made in wood and polycarbonate with solar cells. It is under construction in the harbour of Copenhagen, Denmark, 2007

Excerpts from Learning Poster #006

Learning Site
"Learning Poster #005,
Where is the Sun?
Gleaning the Solar Power"
Print on paper
A1 (59.4 x 84 cm)
2007
Site-specific artwork created
for "Weather Report: Art and
Climate Change,"
Boulder, Colorado

Ellen K. Levy, a New York–based artist and teacher, is immediate past president of the College Art Association. She has played a leading role in airing issues of complex systems and art in her art, publications, and lectures. Levy has exhibited her work widely, both locally and internationally. In 1985 Levy received an arts commission from NASA. The artist was a recipient of an Association International des Critiques D'Art (AICA) award (1995–1996) and later a Distinguished Visiting Fellow of Arts and Sciences at Skidmore College in 1999, a position funded by the Luce Foundation. Levy has had solo exhibitions at the Ezra and Cecile Zilkha Gallery at Wesleyan University and at Michael Steinberg Fine Arts in New York during 2006–2007. Her work was included in the exhibition Petroliana *at the second Moscow Biennale and is part of a traveling exhibition about Mendel, originating at the Field Museum in Chicago.*

Ellen K. Levy

2020 Vision 1 and 2 includes scanned images and text from a brittle, color-aged 1970 issue of the *Whole Earth Catalog.* The counterculture's idealistic agenda was to place self-determination in the hands of the people, providing access to technology that was humanely scaled and ecologically sound. Levy has painted extreme visions over these scanned pages: one of drought eclipsing "Bucky" domes and another of untimely ice calving, violently displacing Arctic pipelines and a radome (plastic housing sheltering the antenna assembly of a radar set). *Sea Currents* highlights images and texts from registered patents to explore the cumulative negative impact of industries on the Arctic. The three panels together offer commentary on the nature of capitalism, innovation, and the changing perceptions of technology's environmental impact.

Ellen K. Levy
"2020 Vision 1," 2007
"2020 Vision 2," 2005
Digital prints, enamel, acrylic
Each 80" x 24"
Courtesy of Michael Steinberg
Fine Art, New York

Iñigo Manglano-Ovalle was born in 1961, in Madrid, Spain, and currently lives and works in Chicago. He received a Bachelor of Arts degree in Art and Art History as well as Latin American and Spanish literature from Williams College in Williamstown, Massachusetts in 1983. In 1989, he completed his Master of Fine Arts degree in sculpture from the School of the Art Institute of Chicago. Described as "one of his generation's most compelling, socially-engaged and celebrated young artists," Manglano-Ovalle has exhibited his work at acclaimed institutions including the Museum of Modern Art and the Whitney Museum of American Art in New York City, as well as Documenta 12, Kassel, Germany. He has received the John D. and Catherine T. MacArthur Foundation fellowship (2001), the Media Arts Award from the Wexner Center for the Arts (1997) and a National Endowment for the Arts fellowship (1995). Manglano-Ovalle is represented by Max Protetch Gallery, New York City.

Iñigo Manglano-Ovalle

Manglano-Ovalle investigates diverse subjects such as technology, climate, immigration, and the global impact of social, political, environmental, and scientific systems. In the process of creating his work, Manglano-Ovalle often collaborates with technical experts across multiple disciplines including engineering, architecture, genomics, and climatology to produce objects that are technically complex, formally captivating, and conceptually engaging.

Iceberg (r11i01) is based on scientific data of an iceberg drifting in the Labrador Sea. This iceberg was scanned with the assistance of the Canadian Hydraulic Center utilizing both radar and sonar. Using data provided by the Center, the artist worked closely with Chicago architect Colin Franzen to create a 25-foot sculpture comprised of thousands of aluminum tubes and rapid-prototyped joints. In selecting the form of an iceberg, Manglano-Ovalle explores the allegorical and other symbolic aspects of this natural phenomenon—the unknown below the surface, the irrepressible and evolving force of nature, a structure non-observant of political borders or boundaries, and the proverbial "tip of the iceberg."

Mapped and produced using similar methods as *Iceberg (r11io1)* but with fewer points of data and in smaller scale, *Black Berg (r39oi02)* encloses a meteorological weather balloon referencing the scientific processes used to understand natural phenomena such as icebergs, clouds, and weather systems. The near perfect sphere of the balloon introduces an additional formal component while also hinting at a primary method of construction and design—the use of triangulation to achieve rigidity and strength—a technique inspired by Buckminster Fuller's geodesic dome.

Iñigo Manglano-Ovalle
"Iceberg (r11i01) Beyond the Irish Sea"
Installation in Liverpool, England, 2005
Courtesy of the artist and Max Protetch Gallery, New York

Patrick Marold grew up in Colorado, cultivating an active relationship between creative endeavors, the landscape and a sensitivity to our habitation as humans. After earning a degree in industrial design from Rhode Island School of Design in 1997, he assisted several artists while resolving and reinforcing his own approach to the landscape and material. Over the years Marold has received various awards and opportunities throughout the world that have proved instrumental in his development as an artist. In 2000, Marold was awarded a Fulbright Fellowship to carry out a year of sculptural responses within Iceland, and an Honorary Residency at the Nantucket Island School of Art. He now resides in Denver with his wife, continually building on his tactile explorations within the urban and rural environments, including installations like the Windmill Project.

Patrick Marold

The Windmill Project reveals the shape and behaviors of air currents passing over the landscape. Each windmill individually responds to the velocity of wind that passes by, generating a relative value of light. As one complete installation of windmills, it becomes a living body of light derivative of the wind's own path and force. The patterns and shapes of wind are translated into momentary displays of blooming and fading bodies of light.

This mechanical system, combined with nature, calls attention to the enormous potential in wind and exhibits wind's beauty and distinct personality.

Patrick Marold
"Windmill Project:
Vail, Colorado"
2700 windmills about
8' high (aluminum wind
cups, generators, electrical
components and LED bulbs,
mounted in polycarbonate
tubes)
Installation in Vail, Colorado
2007-ongoing
Courtesy of Artyard and Peggy
Mangold, Denver, Colorado

Natasha Mayers has supervised more than 500 murals as a touring artist with the Maine Arts Commission since 1975. The painted utility poles in her town depicting local history were featured in Lucy Lippard's book, The Lure of the Local. *She is artist-in-residence for Peace Action Maine, and was a National Endowment for the Arts Millennium Artist, creating community art that explored local views of identity, values, and sense of place, demonstrating how involvement in the arts can improve the quality of community life. In 2005 she received the Arthur Hall Award "for an artist whose work, community service and commitment to their craft inspires others around them to reach to their highest potential." She was awarded the Individual Artist's Fellowship from the Maine Arts Commission, the Artists Projects: New Forms Award from New England Foundation for the Arts, and the Zorach scholarship to the Skowhegan School of Painting and Sculpture.*

Natasha Mayers

For every Fourth of July parade in Whitefield, Maine, Natasha Mayers and her family and friends make a float about a current issue, like global warming, tax cuts for the wealthy, clear cutting, our addiction to oil, and most recently, the disappearance of the honeybees. This motley group of about twenty-four people ages six to eighty always provokes laughter, thought, puzzlement, and sometimes a little hostility. They find that humor and creativity disarm people and get them to pay attention, like images of grown men in diapers trying to lift the national debt or a gas-guzzling SUV that eats protesters. Their floats usually get second or third prize, but never first because they are not about the Liberty Bell or Betsy Ross. Last year's float featured two opposing groups—those who welcome global warming and those who are concerned about it. Some of those in the "global warming welcome committee" included a real estate salesperson selling future oceanfront property (Whitefield is twenty miles from the coast), an alligator hunter, an ice cream vendor, an air conditioner salesperson, a tropical fruit grower, a bottle of sunscreen, and a golf-playing dermatologist. Those concerned about it included a snowplow driver, a cross-country skier, a hockey player, a snowmobiler, and a melting iceberg.

Natasha Mayers
From "Whitefield Welcomes Global Warming/Whitefield Concerned Citizens Urge Action on Global Warming"
"Hot new fashions from L.L.Bean winter collection"
4th of July Parade,
Whitefield, Maine
2006

Jane McMahan is a conceptual artist, painter, educator, and social activist living and working in Colorado. Her work explores ways to visually capture the process of construction, deconstruction, and reconstruction and the moments when they occur simultaneously. Recurring themes in her work include astronomy, biology, meteorology, cultural interactions, and grids. McMahan received her Bachelor of Fine Arts degree from the University of Colorado in 1976. She taught art in the public schools for 12 years before returning full time to her own work in 1998. Selected recent exhibitions include The Drawing Show, *Boulder Museum of Contemporary Art;* Open Studios, *Boulder, Colorado;* Group Show, *Spark Gallery, Denver; and a solo exhibition,* To Change, *at The Dairy Center for the Arts, Boulder.*

Jane McMahan

"And men shall fashion glaciers into greenness and harvest April rivers in the Autumn," reads the inscription under a mural in the Colorado State Capital.

This installation is a metaphor for the challenge of preserving invaluable glaciers from the effects of global climate change. The work carefully preserves an ice specimen from Arapaho Glacier—a watershed owned by the city of Boulder—with human intervention and technology, just as we are attempting to do with the planet. It is estimated that the glacier has receded 100-130 feet since 1960. The work honors the significance of mountain runoff as a major source of water for the Front Range and the Southwest.

The piece of ice is on life support, suggesting the fragile point at which we now find our environment. The ironic twist of this installation is that the sun is the source of energy to keep the ice specimen frozen. Will alternative energy sources be part of the solution for climate change? Just as human and technical intervention is needed for the ice block to survive, scientific and technical knowledge as well as human intervention are needed to maintain the health of the planet.

Jane McMahan
"Arapaho Glacier: What Goes Around Comes Around"
Arapaho Glacier ice, glass, steel, refrigeration equipment, solar panels, batteries, aluminum screen, pump, hose, Boulder Creek water
Ice box 72" x 32" x 32", solar frame 6' x 10' x10'
2007
Site-specific artwork created for "Weather Report: Art and Climate Change,"
Boulder, Colorado

Mary Miss, an artist known for her environmentally-based artwork, lives in New York City. For more than four decades Mary Miss's work has examined the intersection of sculpture, architecture, environmental engineering, and installation art in projects and proposals ranging from riverfront walkways to infrastructure sites. Grounded in the context of place, Miss creates installations that allow the visitor to become aware of the site's history, ecology, and surrounding environment. Permanent installations include Framing Union Square *at the Union Square Subway Station in New York City and a wetlands preservation project in Des Moines, Iowa. Recently she was a lead designer of the collaborative team that won the competition to design the 1,300-acre Orange County Great Park currently being built in Irvine, California.*

Mary Miss

How can the imagination be provoked to envision an event outside our daily experience, one that completely alters the landscape we occupy? How can the predicted flooding of Boulder Creek be made tangible to the residents of this city? A three-dimensional mapping of the floodwaters depth and extent is done at specific locations in the historic center of Boulder. Blue discs are attached to trees along the creek to indicate the estimated potential maximum water level during flood conditions. Various buildings in the floodplain such as the Library, Boulder High School, and City Hall also have high water marks noted on them. Blue marks on the pavement indicate the outer flood plain limits of historic and future floods. Looking from one point to the next, connecting the dots, the level of a flood is no longer abstract. This is one part of a larger study about the flooding of Boulder Creek that would more fully reveal additional aspects of the nature of a flood, the history of flooding in Boulder, and how future floods would affect the city. The installation in Boulder was supervised by Jonathan Feier.

This project was done with the collaboration of retired geologist Peter W. Birkeland, University of Colorado at Boulder, and hydrologist Sheila Murphy, US Geological Survey, assisted by Charles Cross and Phanat Sonemangkhala.

Mary Miss
"Connect the Dots: Mapping the High Water, Hazards and History of Boulder Creek"
Mixed Media
2007
Site-specific artwork created for "Weather Report: Art and Climate Change,"
Boulder, Colorado

Born in Iowa in 1944, Joan Myers has spent a lifetime learning and exploring. After receiving her MA in musicology from Stanford University in 1967, Myers took up photography in the mid-1970s. Her work has been featured in numerous exhibitions and is included in the collections of the Museum of Modern Art, Center for Creative Photography, George Eastman House, and the San Francisco Museum of Modern Art. Monographs of Myer's work include Pie Town Woman *(2001),* Salt Dreams: Land and Water in Low-Down California *(1999),* Whispered Silences: Japanese Americans and WWII *(1996), and* Santiago: Saint of Two Worlds *(1991). In 2002, the National Science Foundation awarded Myers an Antarctic Artists and Writers Grant to photograph the South Pole during the 2002-2003 austral summer. The work was featured in a traveling exhibition,* Wondrous Cold: an Antarctic Journey *which opened at the National Museum of Natural History in Washington, DC in 2006, accompanied by a book published by Smithsonian Books.*

Joan Myers

A National Science Foundation grant gave Myers the opportunity to photograph at McMurdo Station, other surrounding field stations, historic huts, and the South Pole during the 2002–2003 austral summer. Antarctica is an alien outpost, overcast and harsh, colonized by a few thousand scientists and support staff. "Antarctica is also the most mysterious, intriguing, and spectacularly beautiful setting I've ever encountered," Myers says.

It is perhaps ironic that in this most hostile of environments lie clues for the survival of our species at a time of global climate change. Changes in the west Antarctic ice sheet and sea ice have immediate consequences for land and ocean heating, sea level rise, and terrestrial and marine ecosystems.

Joan Myers
"Barne Glacier, Antarctica"
Digital print using Cone
all-black inks on Hahnemühle
paper
12.5" x 34"
2002
Courtesy of the artist and
Andrew Smith Gallery, Santa Fe

Beverly Naidus intertwines the roles of activist, educator, writer, and interdisciplinary artist. Her media range from interactive, site-specific, audio installations to digitally-rendered artist's books. Inspired by lived experience, she has made art about the ecological crisis, healing body hate, the shame and pride of being "other," unemployment, and the nightmares of nuclear war. Early recognition in the alternative margins of the New York City and Los Angeles art worlds brought her many opportunities to exhibit internationally, publish artist's books and articles, visit campuses as a guest artist, and receive significant reviews in contemporary art books and journals. She is a recipient of many university grants as well as the Massachusetts Cultural Council Artist's Grant in Photography (2001). She graduated from Carleton College in Northfield, Minnesota (Bachelor of Arts) and the Nova Scotia College of Art and Design (Master of Fine Arts). She is currently teaching art for social change in an interdisciplinary program at the University of Washington, Tacoma, and lives on Vashon Island with her husband and son.

Beverly Naidus

Culture jamming as a subversive art practice informs this series of digital monoprints. Advertisements from the post-World War II era with their innocent promotions of comfort, safety, and progress have been altered to include the stories of those who either are still living in denial about ecological collapse or whose despair and/or cynicism is palpable. Corporations whose goals of profit outweigh any sense of common good are satirized. Like past projects dealing with environmental illness *(CANARY NOTES: The Personal Politics of Environmental Illness)* and racism *(☑ Other: Breaking Out of the Box)*, this body of work uses painful humor to question our complacency and provoke discussion.

Beverly Naidus
From the series "Brought to You by 'Men Who Plan Beyond Tomorrow'"
"Doom Plastics," 2006
Digital monoprints
8.5" x 11"

Chrissie Orr was born in Scotland and attended the Edinburgh College of Art. She is an artist, animateur, and activist who focuses on developing "an aesthetic around community and site with issues relevant to both." Orr has created innovative community-based projects in Australia, Iran, Turkey, Europe, Mexico, and the United States. She has lectured internationally on her work, especially regarding the Bridge Project *that addressed issues on the border between El Paso, Texas, and Cuidad Juarez, Mexico. She was the founder of the nationally acclaimed* Teen Project *in Santa Fe, New Mexico, receiving recognition from Congress and the National Endowment for the Arts. In 2000, Orr completed a nine-month residency in Georgia as part of the Artists and Communities for the Millennium Project. In 2006 she was the artist in residence at Grand Central Arts Center in Santa Ana, California. She is presently completing two community arts-based projects in Santa Fe, New Mexico.*

Chrissie Orr

"Water Supply in Colorado and the intermountain west has been fraught with uncertainty. Climate change represents another dimension of the uncertainty, and it is an important one."
—Lee Rozaklis, an engineer at Hydrosphere Resource Consultants

"There is a heart beat in Boulder Creek; sometimes it is a double beat. Will this continue?"
—from a conversation with Lee Rozaklis, April 5, 2007

Water is life. We cannot survive without it.

It is predicted that by 2100 the average global temperature will have risen by 4°C.

This warming trend is already affecting snowmelt, run-off, and soil moisture, increasing evaporation rates that produce extremes in drought and flooding. The flow patterns in the watershed will shift; the water will find a new path—either plunging underground or carving new routes and careening away valuable topsoil or totally disappearing and leaving behind only a marked memory. *Earth Drawing* is a poetic, regenerative translation of the predicted change to the patterns of water flow in the Boulder Creek and the effects on the lives of those living in the watershed.

Chrissie Orr
"REMNANTS: Earth Drawing 4"
Recycled Corten steel, mulch, compost, soil, clay, rock, pebbles, seeds, tumeric, and found natural materials
2007
Site-specific artwork created for "Weather Report: Art and Climate Change," Boulder, Colorado

"Earth Drawing"
Watercolor paper, India ink, walnut ink
15" x 22.5"
2007

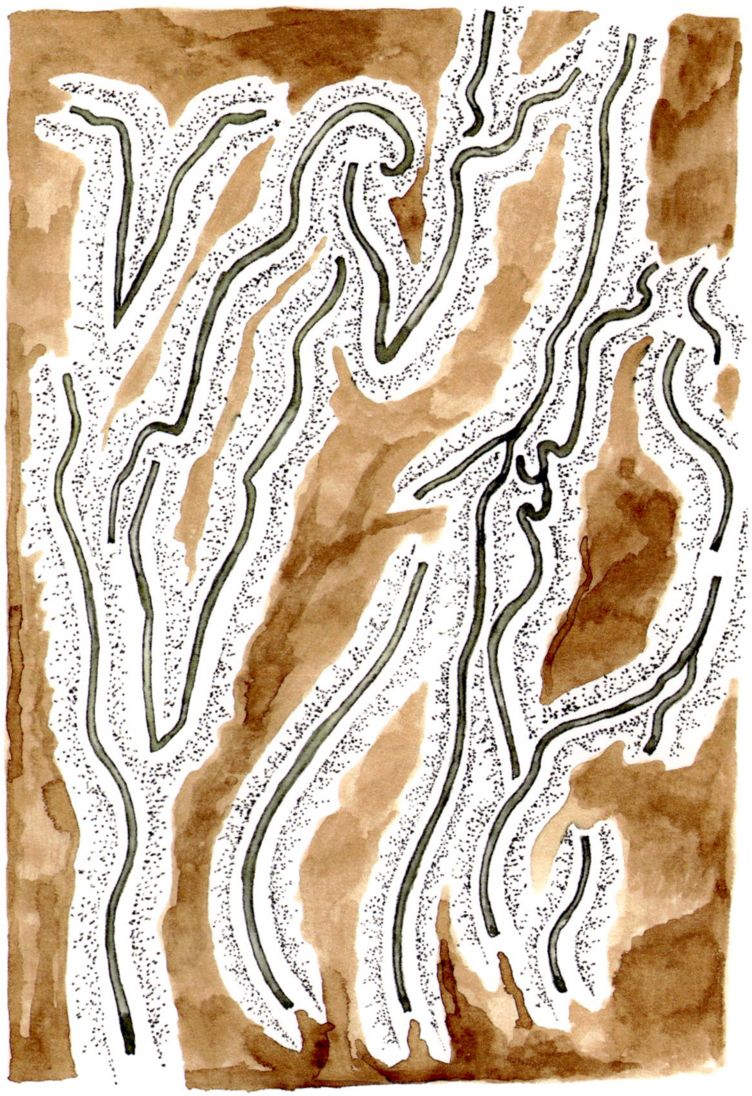

Andrea Polli is a digital artist based in New York City and an associate professor in film and media and integrated media arts at New York's Hunter College. Her work addresses issues related to science and technology and their relationship to contemporary society. Described as "hybrids," her projects are said to reside where science and art intersect. She is currently collaborating with meteorologists and environmental scientists to develop systems for understanding and interpreting climate change through sound. Polli's work has been presented nationally and internationally through exhibitions and performances, and she has been the recipient of numerous awards, residencies and grants, including a grant from UNESCO. In 2007/2008, she will spend two months working with climate scientists in Antarctica as a National Science Foundation polar artist. Polli is currently the co-chair of the Leonardo Education Forum and a founding member and co-chair of the New York Society for Acoustic Ecology.

Andrea Polli

According to NASA climate scientists, a dramatic warming trend in the Arctic may accelerate global climate change. The *N.* installation expresses the isolation and environmental extremes of this remote region and addresses the region's importance to the global ecosystem. *N.* presents a sonification of Arctic weather data from 2003 to 2006 modeled specifically for this project by Dr. Patrick Market, meteorologist and snow and ice specialist from the University of Missouri. The work also incorporates imagery from the National Oceanic and Atmospheric Administration's (NOAA) Arctic research program. *N.* is a collaboration with Joe Gilmore and scientific collaborator Dr. Patrick Market, commissioned by the Lovebytes Festival, audio engineered at Harvestworks, New York City by Ken Babb and supported in part by a PSC-CUNY Research Foundation Grant.

Andrea Polli
"N."
Sound and video projection of
computer-based work
2003-2006

Marjetica Potrč received degrees in architecture and fine arts from the University of Ljubljana, in Slovenia, where she was born. Potrč was the recipient of the Hugo Boss Prize in 2000, given in recognition of an artist whose career "exemplifies a significant new development in contemporary art, one with international cultural reverberations." Her work questions the relationships between parallel urban realities while examining self-sustainability, individual initiative, and empowerment. Her work has been exhibited extensively throughout Europe and the Americas, with solo shows at the Guggenheim Museum in New York (2001), the Max Protetch Gallery in New York (2002 and 2005), the Nordenhake Gallery in Berlin (2003 and 2007), the List Visual Arts Center at the Massachusetts Institute of Technology (2004), and The Curve, Barbican Art Galleries, London (2007). Her many on-site installations include Dry Toilet (Caracas, 2003) and Balcony with Wind Turbine (the Liverpool Biennial, 2004). She is the recipient of numerous grants and awards.

Marjetica Potrč

Boulder in a Time of Global Warming predicts the future of Boulder by enhancing the current state of the city and incorporating the possible consequences of global warming. Anticipating an age of migration and climate upheavals—to protect its culture and to cultivate its land—the city reasserts its territory as the Republic of Boulder. It is an example of future territorialization; it remains small in scale, using both hi-tech and local knowledge to survive. In contrast to the sprawling megalopolis, *Boulder in a Time of Global Warming* functions as a closely-knit network of communities.

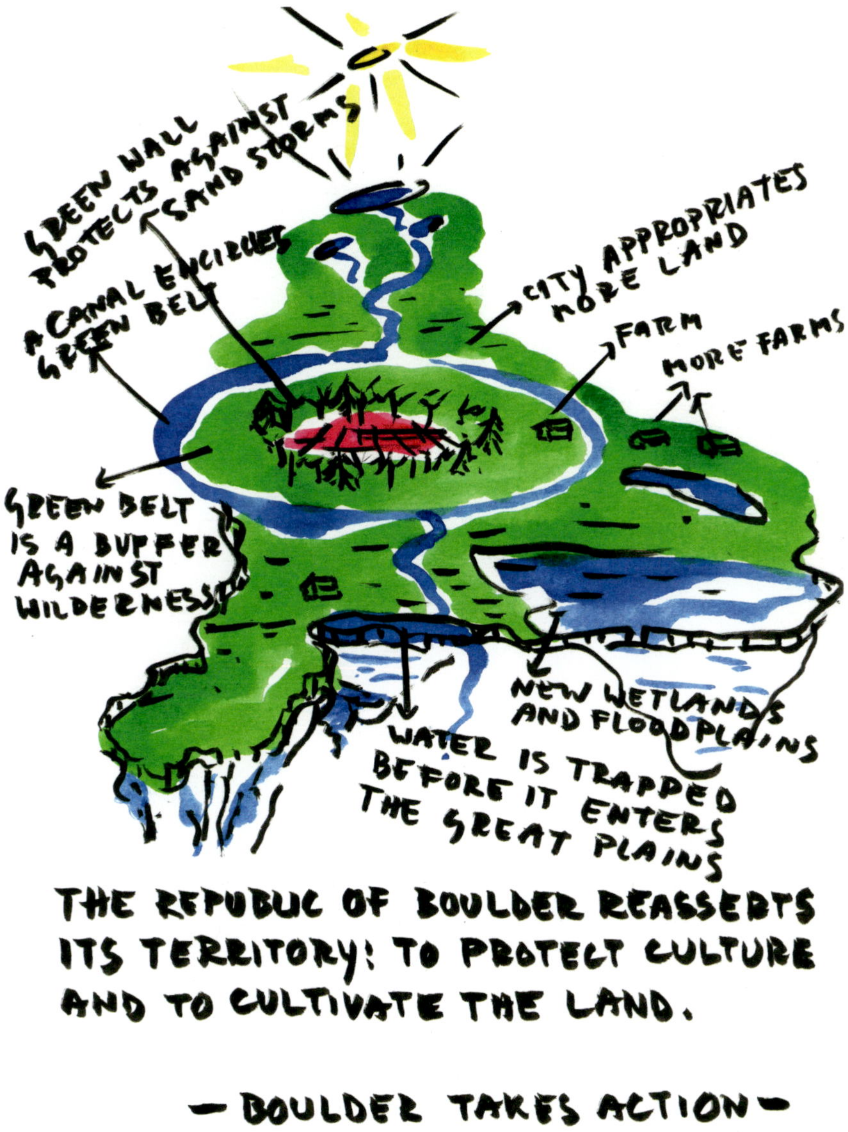

Marjetica Potrč
"Boulder in a Time of
Global Warming"
Ink on paper
8 drawings, each 8.3" x 11.7"
2007
Courtesy the artist
and Max Protetch Gallery, New York

Aviva Rahmani's work has reflected environmental and social concerns throughout her forty-year career. Her projects range from complete landscape restorations to museum venues that reference painting, sound, and photography. Early influences on her work include interdisciplinary classical studies, activism, city planning, and the merging of science with aesthetics. Rahmani's current work reflects her interest in the application of GIS to "explore potential solutions for urban and rural water degradation in large landscapes." Rahmani has taught, lectured, and performed internationally and is the recipient of numerous grants and fellowships, including two from the Nancy H. Gray Foundation for Art in the Environment in 1999 and 2000. She is currently using the Internet "to perform residencies without the international travel that spews jet fuel over the earth's waters. Virtual Cities and Oceans of If *addresses global warming and political conflicts by demonstrating, analyzing and interpreting the local impact of global warming at international real-world sites.*

Aviva Rahmani

Can we envision the ethical implications of our increased carbon loading on our ecological system for people and other species who are unable to move out of harm's way? Can we map the implications and logistics of mass-migrations of species and peoples? All the rules have changed—yet we do not know what the new ones are. We do have historical records that inform our future-casting. Key riparian zones, such as the Mississippi in the wake of Katrina, Bangladesh in relation to the Ganges, and Darfur in relation to the Nile, are examples of the future coming to pass now. Millions of people are forced to adjust to changes faster than biological life is equipped to handle them. The pace of evolutionary time has collapsed. Can small trigger points, in geographically or psychologically identifiable patterns of degradation, be restored to create a significant and measurable mitigating effect? If so, how might we identify and effectuate that change?

This project is presented in collaboration with Dr. Jim White, carbon emissions specialist, Interim Director of INSTAAR, the Institute for Arctic and Alpine Studies, and the past Director of the Environmental Studies program at the University of Colorado at Boulder.

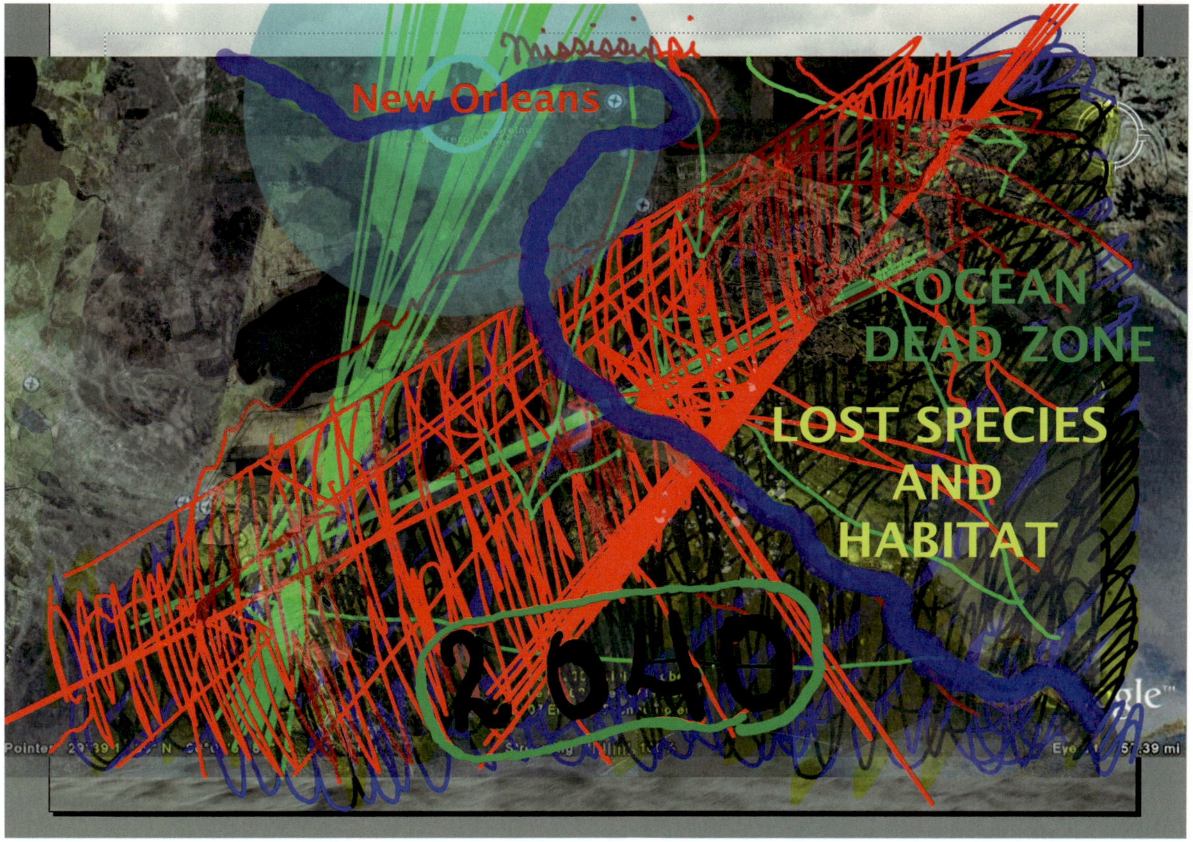

Aviva Rahmani
Detail from "Trigger Points,
Tipping Points"
"Mississippi 2040"
Noting comparisons between the Mississippi in relation to New Orleans, the Ganges in relation to Bangladesh, and the Nile in relation to Darfur in 2040 with the projected impact of global warming, in conversations with Dr. Jim White. Permission pending (for Google Earth Pro satellite mapping).
Digital print
13" x 19"
2007

The participants in Rapid Response were Christina Cobb, Peter Fend, Julia Fischer, and William Meyer III.

Rapid Response

Rapid Response, a dynamic group of artists, designers and architects from New York City, takes art to the streets to highlight the global need for renewable fuel sources. "We are both an agitprop group and a group that develops alternative fueling scenarios we hope will lead to large-scale production of non-fossil-based fuels," they explain. "RAPID is our brand."

Rapid Response's group approach to art echoes the complex flow of money, molecules, and traffic they criticize in their work. *Global Warning* is a global call for action.

Rapid Response
"Global Warning"
Installation at Nikolai Fine Art,
New York, New York
1999
Photo credit: Julia Fischer

Buster Simpson has been active as an artist working in the public since the 1970s. His work ranges from stand-alone sculpture to integrated and/or collaborative works. All of his work incorporates ecological, historical, social, and aesthetic considerations, contextualizing them into the site-specific values of place. Simpson has stated, "I prefer working in public spaces. The complexity of any site is its asset—to build upon, to distill, to reveal its layers of meaning. Process becomes part and parcel to the art of the place." His projects use a variety of media, ranging from rotting wood in Host Analog *to the highly technical in* Brush with Illumination, *and include a $14 million light rail bridge collaborative over the Salt River in Phoenix. Simpson's recent installation,* Incidence, *at the International Museum of Glass, illustrates his use of natural phenomena to engage the space with the viewer. Simpson is based in Seattle.*

Buster Simpson

In the spring of 2002, a group of high school students volunteered for a workshop organized at Lewis and Clark College by the science faculty. The scientist picked a group of four artists involved in environmental issues to work with the thirty students and teams of scientists. The group divided into smaller teams of students and set out to develop a series of agitprops that confronted the issues of global warming. The weeklong workshop included making work, attending a political rally, and taking a journey along a watershed that will be heavily impacted by global warming. The site visits began at the headwaters of Portland, Oregon's drinking water, Mount Hood. White shirts, arms tied together and stretched, were placed on the site as a graphic gesture of resistance and collective effort. They functioned as symbolic bandages for the impact global warming would have on humanity and the environment.

The white shirts marked the receding glaciers of Mount Hood and also resembled prayer flags stretched between two endangered timberline trees. Descending in elevation, additional sites were documented in the watershed, including the parched water reservoirs of the City of Portland. The white shirts also appeared in Portland, confronting the citadels of capitalist shortsightedness and confronting a lineup of automobile exhaust pipes. The shirts concluded their journey at the confluence of the Willamette and Columbia Rivers, where the ocean tidal influences will be impacted by global warming.

The horizontal scroll was the result of the students' collaborative effort to conceptualize and document the impact of climate change. This document was exhibited at PICA in Portland, Oregon in 2002.

"Stretched Shirts" detail

Buster Simpson
"Stretched Shirts"
Performance at Mount Hood, Oregon
Digital print
13" x 69"
2002

Kristine Smock has worked as an artist, teacher, and public art builder for the past twenty years. Her career in public art began around 1989 with the piece Alley Cats*, commissioned by Midyette Architects. Other works include the* Skunk Creek Underpass Project *funded by the City of Boulder, Colorado;* Three Horses*, an interactive piece funded by the Colorado History Museum;* Daydream the Dragon*, an installation in the children's garden at the Boulder Library; and* The Denver*, a piece for the Denver Public Art Committee. In 2003 Smock created pieces for five playgrounds for the Denver Public School District and has directed numerous mural projects at elementary schools in Boulder, Denver, and Lafayette, Colorado. She is currently the assistant director for The Sculpture Trail in Lyons, Colorado. The Trail winds its way through the streets, parks, and back roads of Lyons, facilitating groups of artists, residents, seniors, and students in creating public art for the community.*

Kristine Smock

Born from the earth and sky, the goddess exemplifies three major forms of renewable energy sources: wind, sun, and biomass. Look up to see, hear, and feel the wind and the sun—both powerful endless resources of energy. The sun heats up and powers the lights in homes, businesses, and the tiaras on Our Lady's head. The wind blows through the torch of hope burning a cleaner way of life. The earth is a living biomass that can grow many plants that can be converted to fuel, enabling mobility around the planet. These are the gifts offered to us from the goddess and she suggests we all transform and use alternative fuels soon!

Our Lady is made from welded scrap rebar, clay soil dug from the garden, and advanced technology components, bringing together the new and old. As our beacon of hope, she welcomes us all.

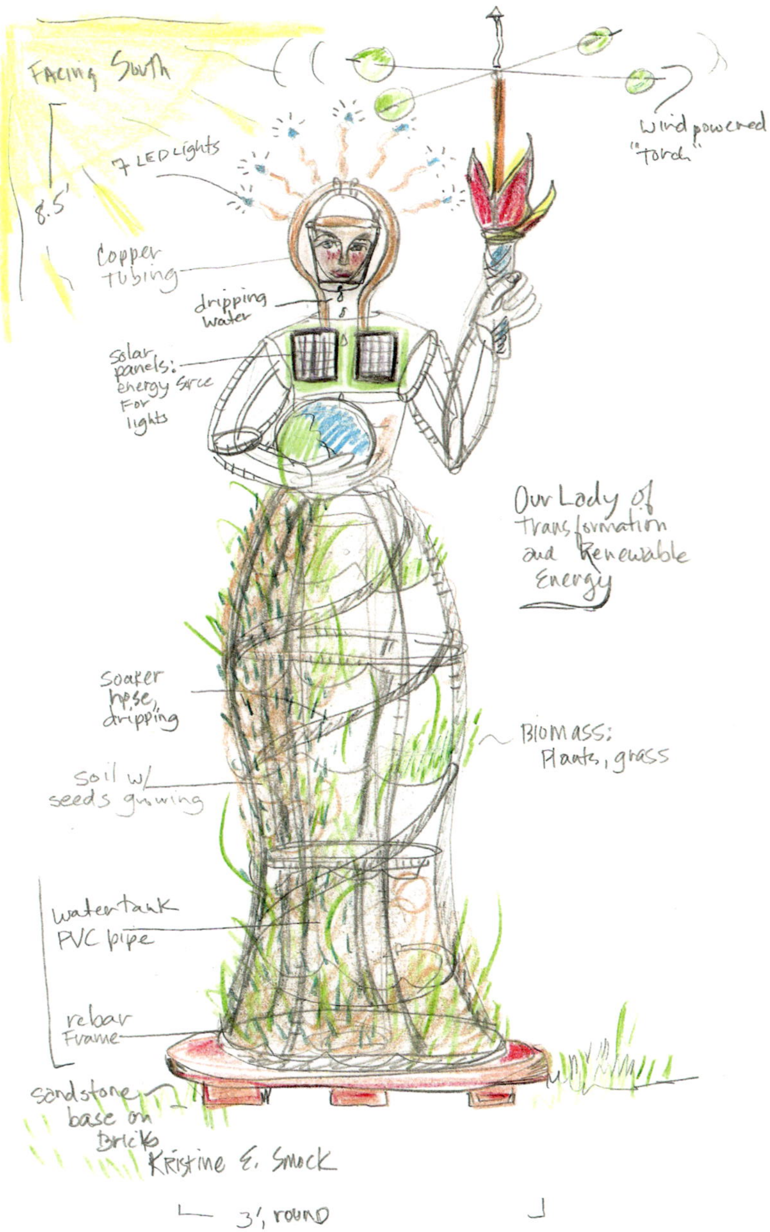

Kristine Smock
"Our Lady of Transformation
and Renewable Energy"
Steel, recycled materials,
solar panels, soil
8' x 3' x 3'
2007
Artist's rendering of site-
specific artwork created
for "Weather Report: Art and
Climate Change,"
Boulder, Colorado

Joel Sternfeld is primarily known for large, color photographs that explore the idea of collective American identity by documenting ordinary people and places. Sternfeld began making color photographs in the 1970s, influenced by the color theory of Johannes Itten and Josef Albers. Using color in highly subtle and sophisticated ways, his work has been described as "wedding the modernist tradition of documentary with the multiple meanings and complexities of postmodern life" and has had a significant influence on a younger generation of color photographers. His awards include two Guggenheim Fellowships, a National Endowment for the Arts Photographers Fellowship, and a Prix de Rome. Recent exhibitions include Stranger Passing *at the San Francisco Museum of Modern Art,* Walking the High Line *at Pace/Wildenstein Gallery in New York City, and* When it Changed, Portraits at the United Nations Climate Change Conference Montreal, November–December of 2005. *Sternfeld currently teaches photography at Sarah Lawrence College in New York.*

Joel Sternfeld

"Future generations are going to wonder about us, the inhabitants of the Earth when the climate began to change. If seas are rising and at the same time drinking water is scarce, they are going to want to know what scientific evidence was before us and what we did in response to it. It is difficult to imagine a time in the past without an image, so I went to Montreal in 2005 to photograph the participants in the eleventh United Nations Framework Convention on Climate Change."
—Joel Sternfeld

Fifteen of those color portraits of participants at the conference form the wall grid *When It Changed*.

The mezzo busto close-ups fit within a tradition of portraiture that dates back to the Renaissance but the anxiety on each face seems like something only modernity and the advent of ecological catastrophe could produce.

By the title *When It Changed*, Sternfeld may also be referring to a hopeful turning point. In the year and a half since the conference, there has been increased global awareness of climate change and positive action to confront humanity's greatest challenge.

As a photograph-based artwork, *When It Changed* represents an extension of traditional documentary practice of recording important moments in human history.

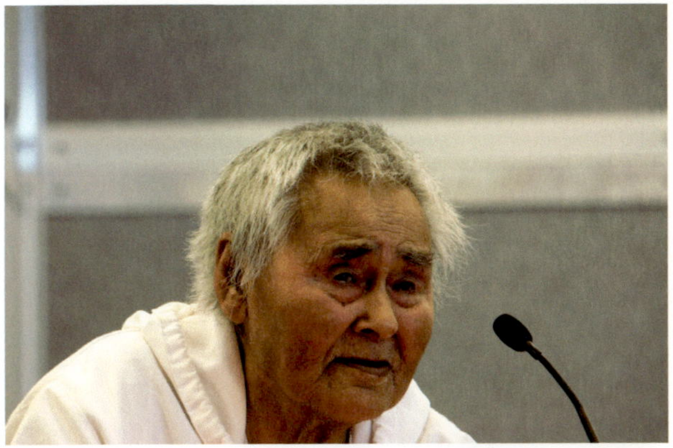

Joel Sternfeld
"When It Changed"
Portraits Made at the United Nations Framework Convention on Climate Change, Montreal, Canada, November/December, 2005
Pigment prints
Overall 141.675" x 59", made up of a grid of 15 portraits, each 16.166" x 24.25"
2005
Courtesy of the artist and Luhring Augustine, New York

Since the late 1960s, Mierle Laderman Ukeles has developed performance and installation work integrating feminism, environmentalism, and labor activism, calling it "Maintenance Art." Her works are inspired by the New York City Comprehensive Plan's mission to support dual systems—development and maintenance—and the works manifest these systems as the embodiment of the opposing drives of freedom and necessity. She has been the New York City Department of Sanitation's unsalaried Artist in Residence for 30 years and is also the Artist of Fresh Kills, formerly the largest landfill on earth. Ukeles has received multiple awards from the National Endowment for the Arts and support from the Guggenheim, Andy Warhol, Joan Mitchell, and Anonymous Was a Woman Foundations. Current exhibitions include a one-person show at the Armory Art Fair in New York City, WACK! Art & the Feminist Revolution *at the Los Angeles Museum of Contemporary Art, and the Sharjah Biennial 8, United Arab Emirates. She is represented by Ronald Feldman Fine Arts in New York.*

Mierle Laderman Ukeles with Meg Duguid

"Before Navajo elders make important decisions, they insist on asking how each decision will affect not only themselves but rather the seven generations that follow. If it might harm those even in the seventh generation forward into the future, they will not allow themselves to make the decision in their own time.

Our culture sees forward only to the end of our own noses. We haven't taken care or we don't care. Now the whole planet is in danger.

I am 'Savta Ima': Grandmother. What kind of safe thriving world will I and my culture leave for my precious grandchildren? They look at me with utter trust in their eyes."

—Mierle Laderman Ukeles,
New York City, 2007

Mierle Laderman Ukeles
with Meg Duguid
"Dear Children,
Can you forgive us?
And by the way...."
2007

A hybrid artwork in two parts. The first part, an edition of 1000 is an open letter to children from the artist—on one half of an 8.5 x 11" page—asking their forgiveness for her generation leaving such a messed up environment that the children will inherit.

The second part is participatory. At the invitation of the artist, a child or young person is asked to help out, by filling in the remaining blank half-sheet with ideas in words or drawings that he or she may have for fixing the environment.

Once filled in, each sheet in the edition becomes an individual piece and each individual artwork becomes complete.

Individual artworks are to be shown together with many others in a public space. Cooperation and collaboration are essential.

Site-specific artwork created for "Weather Report: Art and Climate Change," Boulder, Colorado.

Melanie Walker has an exhibition history that spans more than thirty years. She received a Bachelor of Arts from San Francisco State University and a Master of Fine Arts from Florida State University. She is a mixed-media artist invested in ideas. Her approach to materials includes digital and photographic media, sound, sculpture, installation, and public art. She is presently a professor of art at the University of Colorado at Boulder. George Peters has worked with aerial forms and wind art using installation sculpture, kites, mobiles, wind sound sculpture, and banner works. He has worked in various studio art fields as a visual artist, sculptor, kite craftsman, fiber artist, environmental and installation sculptor, graphic artist, illustrator, architectural modeler, and theater artist.

Peters' and Walker's collaborative works have merged their installation aesthetic and environmental concerns into a third voice of collective endeavors. The collaborative works speak to aspects of atmosphere, light, wind, and the visible and invisible world.

Melanie Walker & George Peters

We live in a river of wind and sun, while underground we extract coal and other fossil fuels to make energy. Gasses emitted from burning these fuels affect the atmosphere. Often solutions to problems lie not in the depths or the distances but within our reach. *Coal Warm Memorial* suggests replacing fossil fuel and coal dependency with the humming of wind power.

The wind sculpture is an experiential work creating a tunnel of sound and light. Made from wooden "power poles," the installation is connected to the ground via cable tension and rocks used as counterweights. Stretched overhead is a series of wind-driven ribbons that emit sound activated by the wind.

Melanie Walker
and George Peters
"Coal Warm Memorial"
Wind installation
2007
Site-specific artwork created
for "Weather Report: Art and
Climate Change,"
Boulder, Colorado

Ruth Wallen is a multimedia artist whose work is dedicated to encouraging dialogue about ecological and social issues. Her multilayered installations and performances have been exhibited at Franklin Furnace in Brooklyn, New York; New Langton Arts and the Exploratorium in San Francisco; Sushi Performance and Visual Arts in San Diego; CEPA Gallery in Buffalo, New York; and numerous national and international group exhibitions. Public installations include interactive "nature walks" at Carmel Mountain, the San Bernardino Children's Forest and Tijuana River Estuary. Additionally, Wallen was a founding member of the artists' collective Las Comadres. Publications about contemporary art include essays in LEONARDO, Exposure, High Performance, TIKKUN, Women's Studies and With Other Eyes: Race, Gender, and Visual Culture. *Wallen is on the faculty of the interdisciplinary arts Master of Fine Arts program at Goddard College in Vermont. In 1995–1996 she was a Fulbright lecturer at the Autonomous University of Baja California, Tijuana.*

Ruth Wallen

Construction and operation of buildings account for nearly half of US energy consumption. Sprawling suburbs and increasingly large single-family homes are glaring examples of unsustainable development. *Preserving Paradise* explores the competing ideals informing suburban expansion in San Diego. Postcards include text and images that chronicle the contrasts between master plans that envision higher density, public transportation, and small commercial centers; environmentalists fighting to preserve sensitive habitat and encourage construction; and developers committed to building the single-family homes they think consumers desire. A second set of postcards asks viewers to contribute their ideas regarding actions necessary to create a survivable future.

Pacific Highlands Ranch
Patio of Model Home

What does choice mean? The real estate industry does all of these marketing studies that say that people want all of these things, but are people really given another choice? Until people learn that there's an alternative and they get a chance to experience it, what they want is what's been marketed. Are they really going in and telling them, well, you can have a much more compact neighborhood with all of these benefits, or are they just saying, do you want a family room, a great room or a third bathroom?
-Bernie Turgeon, SD City Planner

Preserving Paradise
©Ruth Wallen

Ruth Wallen
"Preserving Paradise"
Digital photographs and mixed media
44 postcards 4" x 6", 2 wall panels 13" x 19", postcard stand 68" x 11.5" x 36"
2007
Photo Credit: Ruth Wallen

Sherry Wiggins received her Bachelor of Fine Arts in Sculpture in 1988 and her Master of Fine Arts in Sculpture in 2005—both from the University of Colorado. Living and working in Colorado, Wiggins works as a new media artist, sculptor, public artist, and cultural organizer. Her artwork is reflective and participatory, centered around explorations of environmental, political, and women's issues. Wiggins invites participants and viewers to examine themselves in specific roles and situations through the use of narrative, interrogative, and physical process. Exhibitions and commissions include the Museum of Contemporary Art in Denver, the Khalil Sakakini Cultural Center in Ramallah (Occupied Territories of Palestine), the Universidad Nacional de Colombia in Leticia (Colombia); and the Denver International Airport. Wiggins is a founding member of 6+: a women's art collective that facilitates exhibitions, publications, and workshops among women artists internationally.

Sherry Wiggins

The intention of the art project *Carbon Portraits* is to understand and personalize our own participation in global climate change. Carbon emission, which is the direct result of fossil fuel consumption, is the major factor in worldwide climate change.

General information about carbon emission is utilized in the work, along with calculations and interviews with individual community members. Gas and electricity bills, car mileage, airplane trips, and the consumption of manufactured products are examined and CO_2 emissions are quantified.

These narratives, data, and calculations are the materials that generate *Carbon Portraits.* The project is a series of billboard-style posters that illustrate how general policy and individual lifestyles and choices affect carbon emissions. Participants and viewers are asked to reflect on their own energy consumption patterns and contemplate ways to change the behaviors that contribute to global warming.

Suzie - CO2 Emmissions (2006)

Household Electricity 4865 Kilowatt Hours (offset by windsource)	0 lbs
Household Natural Gas 307 Therms	330 lbs
Automobile Trips 2003 Prius - 3500 miles	1,800 lbs
Airplane Trips 12 roundtrips	25,200 lbs

Are there ways in which you are willing to change your behavior to reduce your personal carbon emissions?

"We have changed our behavior at the margins of our energy use – we drive an energy efficient car, we live in a small apartment downtown and walk a lot, and we offset our electrical consumption by buying wind power. The main place where we are unwilling to change our behavior has to do with travel. We are retired and we are still of relatively sound mind and body and we can still take these trips and that's not going to be true for very much longer. Traveling is a great source of pleasure for us."

Sherry Wiggins
"Carbon Portraits"
3 billboards; digital prints
mounted on plywood
Each 3' x 6'
2007
Site-specific artwork created
for "Weather Report: Art and
Climate Change,"
Boulder, Colorado

The Yes Men are a group who use any means necessary to agree their way into the fortified compounds of commerce and then smuggle out the stories of their undercover escapades to provide a public glimpse at the behind-the-scenes world of big business. The stories are often both shocking and hilarious. Author Naomi Klein called them "the Jonathan Swift of the Jackass generation." The Yes Men have impersonated World Trade Organization, Dow Chemical Corporation and Bush administration spokesmen on television and at business conferences around the world. They do this (a) in order to demonstrate some of the mechanisms that keep bad people and ideas in power, and (b) because it's absurdly fun. Their main goal is to focus attention on the dangers of economic policies that place the rights of capital before the needs of people and the environment.

The Yes Men

On June 14, 2007, Yes Men Andy Bichlbaum and Mike Bonnano posed as ExxonMobil and National Petroleum Council (NPC) representatives. They delivered a keynote speech to 300 oilmen at GO-EXPO, Canada's largest oil conference in Calgary, Alberta. In the speech, the "NPC rep" (Bichlbaum) announced that current US and Canadian energy policies (notably carbon-intensive exploitation of Alberta's oil sands and the development of liquid coal) are increasing the chances of huge global calamities. But he reassured the audience that in the worst-case scenario, the oil industry could "keep fuel flowing" by transforming the billions of people who die into oil. The audience then lit "commemorative candles" supposedly made of Vivoleum obtained from the flesh of an "Exxon janitor" who died as a result of cleaning up a toxic spill.

In the words of Bonanno, "If our idea of energy security is to increase the chances of climate calamity, we have a very funny sense of what security really is. While ExxonMobil continues to post record profits, they use their money to persuade governments to do nothing about climate change. This is a crime against humanity."

The Yes Men
"Exxon Vivoleum"
Video
2007

Shai Zakai, who holds a Master of Arts in Art and Environmental Policy, is a photographer, multimedia artist, and green activist. In addition to her work in her photography and eco-art studio, she lectures extensively and has exhibited and taught worldwide. Her artwork appears in private and public collections. Zakai is the director and founder of The Israeli Forum for Ecological Art *and has initiated several community green movements and eco-art projects in Israel. She was chosen to represent Israel with* Faces and Facets, *her album and a traveling photo exhibition in the United States in 1998. Zakai also has presented her work in international conferences such as* Art, Community and Environment *(Nevada);* Art Culture Nature *(Arizona);* Earth-Summit *(Johannesburg); and the* World Water Forum *(Kyoto 2003). Zakai received the Artist-Teacher's Prize for 2000–2001 from the Ministry of Science and Culture, as well as the International Photography Biennale prize. Her pioneering stream reclamation project,* Concrete Creek *(1999–2002), has been exhibited internationally.*

Shai Zakai

Data compiled from the American Meteorological Society report prepared by the Intergovernmental Panel for Climate Change and a current Israeli scientific report commissioned by the Ministry of The Environment's Protection shows that thus far scientists are focusing on human adaptation to climate change. It indicates, among other conclusions, that international agreements cannot be the only reliable solution to global climate change. Recommendations include investing in new technologies, and expanding scientific research, while the gap between human and nature compatibility ever widens. In these reports, there is no mention of the effects on the cultural and social components of society.

The artist contemplates if she were able to write a climate change report, she would begin with studies of ancient cultures. For this "self-invited report" she conducts a photographic and research survey that highlights the long gone wisdom of the ancients, specifically regarding agriculture and water issues affected by extreme weather situations. Adopting historical ideas may help reduce weather changes, hopefully raising personal and cultural awareness of climate change issues.

The Nabataeas understood the desert climate very well. They collected rainwater with just a simple relocation system of the mountains' stone. The Tulilot allowed rainwater to flow into water reservoirs and streams as well as fertilizing the soil.

Shai Zakai
"Adaptation & Mitigation—
Learning from the Ancients"
Still images from video
Each 20" x 13"
2006-7

Original artwork: fourteen
metal and recycled wood
"Landscape Units," seven
outdoors and seven indoors
"Erosion" from "Genus Loci"
Color photograph
20" x 13"
2004

Exhibition Checklist

Kim Abeles

"Presidential Commemorative Smog Plate (Herbert Hoover in 25 Days of Smog)"
"Presidential Commemorative Smog Plate (Franklin Roosevelt in 4 Days of Smog)"
"Presidential Commemorative Smog Plate (James Carter in 8 Days of Smog)"
"Presidential Commemorative Smog Plate (Ronald Reagan in 40 Days of Smog)"
"Presidential Commemorative Smog Plate (George Bush in 40 Days of Smog)"
Smog, porcelain, gold
Each 10.5" diameter
1992

Lillian Ball

"66 Degrees, 32 minutes, 50 Years"
Map morph animations projected on ice (shown in photo) or ice water (as currently exhibited)
2007

Subhankar Banerjee

"Caribou Migration I," from the series "Oil and the Caribou"
Digital chromogenic print face-mounted to Plexiglas and framed with wood
86" x 68"
2002
Courtesy of the artist

"Gwich'in and the Caribou," from the series "Blood in my Meat"
Charlie Swaney and Jimmy John; Danny Gimmel and Rocky John; Jimmy John; Danny Gimmel
Pigmented inkjet print laminated and mounted on sintra
Polyptych with 4 prints, each 31" x 39"
2007
Courtesy of the artist

"Bear Den II," from the series "Oil and the Caribou"
Digital chromogenic print face-mounted to Plexiglas and framed with wood
68" x 86"
2002
Courtesy of the artist

"Brant and Snow Geese with Chicks," from the series "Oil and the Geese"
Digital chromogenic print face-mounted to Plexiglas and framed with wood
68" x 86"
2006
Courtesy of the artist

"Known and Unknown Tracks," from the series "Oil and the Geese"
Pigmented inkjet print laminated and mounted on Plexiglas and framed with wood
31" x 41"
2006
Courtesy of the artist

"Exposed Coffin," from the series "Oil and the Caribou"
Pigmented inkjet print laminated and mounted on Plexiglas and framed with wood
31" x 41"
2006
Courtesy of the artist

"Caribou Skeleton," from the series "Oil and the Caribou"
Pigmented inkjet print laminated and mounted on Plexiglas and framed with wood
31" x 41"
2006
Courtesy of the artist

"Beluga Whales with Calves," from the series "Oil and the Whale"
Pigmented inkjet print laminated and mounted on Plexiglas and framed with wood
31" x 41"
2006
Courtesy of the artist

"Caribou Crossing Utukok River," from the series "Coal and the Caribou"
Pigmented inkjet print laminated and mounted on Plexiglas and framed with wood
31" x 41"
2006
Courtesy of the artist

Iain Baxter&

"Animal Preserve #8"
Mixed media installation
26" x 9" x 4"
1999-2007
Courtesy of the artist and Corkin Gallery, Toronto, Canada

"Killer Still-life #5"
Mixed media installation
Variable
1993
Courtesy of the artist and Corkin Gallery, Toronto, Canada

Bobbe Besold

"The Thirteenth Tipping Point"
Performance
2007
Site-specific artwork created for "Weather Report: Art and Climate Change," Boulder, Colorado

Cape Farewell

Photographic documentation from the 2001 expeditions

Mary Ellen Carroll

"indestructible language"
Photographic documentation of installation in Jersey City, New Jersey
Original artwork: neon signs
900' x 8'
2007

Center for Land Use Interpretation (CLUI)

"Dauphin Island, Alabama, November, 2005"
Center for Land Use Interpretation Photo Archive
Inkjet prints on paper
Series of 20 photographs, each 16" x 20"
2005

Brian Collier

"The Pika Alarm"
Audio-rigged sculptures: fiberglass, steel, wood, horn speaker, motion-triggered audio player, informational postcards, website: pikalarm.net
2007
Site-specific artwork created for "Weather Report: Art and Climate Change," Boulder, Colorado

Xavier Cortada

Flag from "The Markers" series
"1956 South Pole Marker (90°S)"
1956: Construction crews arrive at the South Pole. On October 31, 1956, Lt. Shinn landed the first plane, the Que Sera Sera, at the South Pole. Three weeks later, on November 20, 1956, the first South Pole station construction crew arrived. Then, on January 4, 1957, the Navy Seabees crew turned the completed station over to a team of nine scientists, nine support professionals (e.g., a doctor, a cook, etc.) and a dog, who wintered over and officially opened the base to scientific exploration. Fifty years later, on January 4, 2007, Cortada planted "The Markers" at the South Pole.
Cloth and permanent ink marker
32" x 48"
2007

Flag from "The Markers" series
"1979 South Pole Marker (40°01'N, 105°16'W)"
1979: Ten independent living centers were founded across the US for persons with disabilities, including one in Boulder. This flag marked the location where the South Pole stood in 1979 at that historic moment.
Cloth and permanent ink marker
32" x 48"
2007

Flag from "The Markers" series
"2006 South Pole Marker (27°08'S, 109°22'W)"
2006: Scientists discover new species. Near Easter Island, Census of Marine Life researchers discovered a crab so unusual it warranted a whole new family designation, Kiwaidae. The discovery also added a new genus, Kiwa, named for the mythological Polynesian goddess of shellfish. Its furry appearance inspired its species name, hirsute, which means "hairy" in Latin.
Cloth and permanent ink marker
32" x 48"
2007

Video (with the participation of Bill Bilowit) from "The Markers" series
NTSC DVD-Video Disc/16:9 Anamorphic Format
Duration: 5m04s, Auto-play Looped Program
Edition of 5
2007

Soundscape (with Juan Carlos Espinosa) for "The Longitudinal Installation"
Digital music
Duration: 11m02s, Autoplay Looped Program
2007

Photographic documentation of "The Longitudinal Installation"
"The 150-year Journey"
"The Markers"
"Endangered World"
Digital prints
8" x 12" each
2007

Gayle Crites

"Kyoto Protocol"
Brush and ink on Japanese handmade paper
Diptych, 16" x 44"
2006
Courtesy of 15th Street Studio, Boulder, Colorado

"Biota Futurama—Recipe for Bio-Fuel"
Brush and ink with vegetable paper collage
27" x 19"
2007
Courtesy of 15th Street Studio, Boulder, Colorado

"Polar Bear Wreath"
Brush and ink on Japanese Okawara with wax, pastel and embroidery
70" x 39"
2007
Courtesy of 15th Street Studio, Boulder, Colorado

Agnes Denes

(continued from pg. 42)
"Tree Mountain, conceived in 1982, affirms humanity's commitment to the future well being of ecological, social and cultural life on the planet. It is designed to unite the human intellect with the majesty of nature. Tree Mountain was dedicated in June 1996, by the President of Finland, other heads of state, and people from everywhere." — Aira Kalela, Ministry of Environment. Finland

"Tree Mountain - A Living Time Capsule—11,000 Trees, 11,000 People, 400 Years," 1992-96, (420 x 270 x 28 meters), Ylöjärvi, Finland.
Tree Mountain is to be surrounded by two rows of white birches as the green pines become taller.

"Tree Mountain - A Living Time Capsule—11,000 Trees, 11,000 People, 400 Years"
Top image: View of original design drawing from 1983, 35" x 96"
Bottom image: Winter view from 2001 of actual mountain/forest with the mathematical pattern beginning to show
Color photograph on foam-board
9" x 24"
1983, 2001

"Tree Mountain - A Living Time Capsule—11,000 Trees, 11,000 People, 400 Years"
Aerial view, of the mathematical pattern into which the eleven thousand trees were planted.
Color photograph on foam-board
24" x 34"
1995-ongoing

"Tree Mountain - A Living Time Capsule—11,000 Trees, 11,000 People, 400 Years"
People planting their trees on the mountain
Color photograph on foam-board
20" x 24"
1996

"Tree Mountain - A Living Time Capsule—11,000 Trees, 11,000 People, 400 Years"
Summer view from 2003 (detail of forest)
Color photograph on foam-board
24" x 35"
2003

"Forest for Australia"
6000 trees of endangered species, with varying heights at maturity, were planted into five spirals by the artist, creating a step pyramid for each spiral when the trees are full gown. The trees help alleviate serious land erosion and desertification threatening Australia.

"Forest for Australia"
Altoona Treatment Plant, Melbourne, Australia. 400 meters x 80 meters (rendering view of the forest when trees are fully grown)
Color photograph
11" x 14"
1998

"GUT—Grand Unification Theory"
Color photograph
11" x 14"
2002

"Pascal's Perfect Probability Pyramid & the People Paradox—The Predicament (PPPPPPP)"
Original single print etching
T.C.P. 1/1,
30" x 41.5"
1981

"Masterplan—Nieuwe Hollandse Waterlinie"
Twenty-five year Masterplan to bring into prominence and environmentally sustainable the 85 km long defense line dotted with 70 forts built from the 16th to the mid-19th centuries in the center region of the Netherlands: introducing wildlife preserves, bicycle routes, adding 100,000 trees that hold the land from erosion, windmills to pump water and generate electricity, etc. Masterplan includes water and flood management due to weather changes, urban planning, historical preservation and restoration, landscaping, tourism, and community participation.
Five maps
40" x 9" each
2000—

"Crystal Fort"—"Masterplan: Nieuwe Hollandse Waterlinie"
Original design of a full size fortress made of glass, to be built along the Waterlinie as a major tourist attraction to create revenues for the reclamation plans laid out in the 25-year Masterplan. (120' x 120' x 60')
Color photographs
20" x 30"
2000—

Steven Deo

From the series "bare trees"
"heart wood"
"tree of life"
"tree of origin"
"tree of knowledge"
Wood, encyclopedia pulp
Each 72" x 48" x 48"
2007
Photo credit: Gene Ogami

Rebecca DiDomenico

"Intentional Mutation"
Transparency emulsions mounted on Plexiglas, glass marbles
11' x 42" x 2.5"
2007

Futurefarmers (Amy Franceschini and Michael Swaine)

"The Unfinished Journey of Carl Linnaeus"
Performance, video, sculpture, drawing
2007
Site-specific artwork created for "Weather Report: Art and Climate Change," Boulder, Colorado

From "The Unfinished Journey of Carl Linnaeus"
"Starting Points: 1. Inside Research Vessel and 2. Researchers' Backpacks"
Collage/Drawing
11" x 14"
2007

From "The Unfinished Journey of Carl Linnaeus"
"Starting Points"
Pencil on paper
11" x 14"
2007

Bill Gilbert

"Bough"
Digital print on Plexiglas
40" x 40"
2007
Proposed site-specific artwork created for "Weather Report: Art and Climate Change," Boulder, Colorado

"For John Wesley Powell: Attempts to Walk the Grid, Gloating Island, September 5, 2006. One hour in each cardinal direction. Orientation: True north."
Digital print on Plexiglas
40" x 40"
2006
Courtesy of the artist

"For John Wesley Powell: Attempts to Walk the Grid, Sand Canyon, September 27, 2006. Walk one hour in each cardinal direction. Orientation: Magnetic north."
Digital print on Plexiglas
40" x 40"
2006
Courtesy of the artist

"For John Wesley Powell: Attempts to Walk the Grid, Good Hope Bay, September 9, 2005. Walk one hour in each cardinal direction. Orientation: True north."
Digital print on Plexiglas
40" x 40"
2005
Courtesy of the artist

Isabella Gonzales

"2¢ worth...your participation is requested"
Mixed media installation
Variable
2007

Green Fabrication

(courtesy of Erik Sommerfeld, Professor, University of Colorado, College of Architecture and Planning)
Students' work includes:

Bill Daher
"Square Chair"
Poplar wood veneer, low-VOC paint, natural cork, post-consumer waste aluminum
Prototype
2006

Megan Haughey
"Resting Place"
3form ecoresin
Prototype
2006

Kellen Schauermann
"Schauermann Worktop"
Powder-coated recycled steel, bamboo plywood
Prototype
2006

Helen Mayer Harrison and Newton Harrison
"The Mountain in the Greenhouse"
Video
2001

Judit Hersko
"Seven Days of Dissolution"
Seven acrylic tanks, seven black sea fans with Capiz shell sculptures of human heart and lung dissolving, one carbon dioxide tank
Acrylic tanks each 12" x 12"
2006
Courtesy of the artist
Photo credit: Robert Steinberger

Lynne Hull
"Wildlife Warning: Global Warming"
Mixed media
18" x 18"
2007
Site-specific artwork created for "Weather Report: Art and Climate Change," Boulder, Colorado

"Cool Choice"
Stickers
Digital print and bumper sticker technology
4" x 6"
2007

Pierre Huyghe
"A Journey That Wasn't"
Presented in documentation format
Original artwork: Super 16mm film and HD video transferred to HD video, color, sound
Duration: 21m 41s
2005
Courtesy of the artist and Marian Goodman Gallery

Basia Irland
"The Gathering of Waters: Boulder Creek, Continental Divide to Confluence"
2007
Site-specific artwork created for "Weather Report: Art and Climate Change," Boulder, Colorado

"Boulder Creek Backpack"
47 glass vials (one for each mile of the creek), recycled car inner tubes, aspen branch, and sticks cut by beaver.
The backpack repository contains objects from the four-month project, "The Gathering of Waters: Boulder Creek, Continental Divide to Confluence," including the logbook, canteen, video documentary, water samples, hydrology reports, and Boulder Creek watershed maps.
Photomontage of backpack being worn
23.5" x 50" x 3"
2007

"Melting Hydrograph"
Carved wooden book, parched earth, melting snow, USGS hydrograph for stream-flow in Boulder Creek at North 75th Street, April 2005 to January 2007.
9" x 18" x 3" and 8.5" x 23" x 2"
2007

"A Gathering of Waters: Boulder Creek, Continental Divide to Confluence"
Directed and filmed by Basia Irland in collaboration with John Davis, Jessica Damon, Curry Rosato, Dr. Jason Neff, Rick Silva, Talice Lee, and numerous Colorado participants. Original song, "Ballad of Boulder Creek," by Fergus Stone.
Video documentary
2007

"receding/reseeding"
250-pounds of ice, frozen from water gathered by project participants; small stones; columbine, blue spruce and mountain maple seeds.
Photographic documentation
20" x 40" x 10"
2007

Patricia Johanson
"Sculpture Garden: Shelter"
Pencil, colored pencil, lipstick
8.5" x 11"
1969

"Building that Cleans Its Own Water"
Pencil, colored pencil
8.5" x 11"
1969

"Building that Collects Water"
Pencil, colored pencil, pastel
8.5" x 11"
1969

"Garden-Cities: Aerial Highway"
Pencil, colored pencil
8.5" x 11"
1969

"Living Apartment Houses"
Pencil, colored pencil
8.5" x 11"
1969

"Cabbage City"
Pencil, colored pencil
8.5" x 11"
1969

"Line Garden: The Breath of Trees"
Pencil, colored pencil
8.5" x 11"
1969

"Air"
Ink and acrylic on vellum
18" x 66"
2007

"Water"
Ink and acrylic on vellum
18" x 66"
2007

"Earth"
Ink and acrylic on vellum
18" x 66"
2007

Chris Jordan
From the series "Running the Numbers: An American Self-Portrait" (2006-2007)
"Plastic Bottles"
Depicts two million plastic beverage bottles, the number used in the US every five minutes.
Archival inkjet prints
60" x 120"
2007

"Jet Trails"
Depicts 11,000 jet trails, equal to the number of commercial flights in the US every eight hours.
Archival inkjet prints
60" x 96"
2007

"Denali Denial"
Depicts 24,000 logos from the GMC Yukon Denali, equal to six weeks of sales of that model SUV in 2004.
Archival inkjet prints
60" x 75"
2006

"Office Paper"
Depicts 30,000 reams of office paper, or 15 million sheets, equal to the amount of office paper used in the US every five minutes.
Archival inkjet prints
60" x 87"
2007

"Jeep Liberty"
Depicts 200,000 Jeep Liberty SUVs, equal to the number of that model sold in the US every year since the Liberty was introduced in 2004.
Archival inkjet prints
88" x 132" in two panels
2006

Marguerite Kahrl
"Homeland Insecurity"
Artificial hemp plants, planters, subliminal sound installation
242" x 88" x 90"
2007
Site-specific artwork created for "Weather Report: Art and Climate Change," Boulder, Colorado

"Study for Homeland Insecurity (a)"
Giclee print
11" x 17"
2007
Courtesy of the artist

Janet Koenig and Greg Sholette
"Cannibal Tech"
Inkjet poster, recyclable computer parts
Each 4' x 4' x 8'
2007
Site-specific artwork created for "Weather Report: Art and Climate Change," Boulder, Colorado

Eve Andrée Laramée
"Pretty Vacant"
4-color ink on newspaper stock
Each 11.625" x 10.75"
2007
Site-specific artwork created for "Weather Report: Art and Climate Change," Boulder, Colorado

Learning Site
"Learning Poster #005, Where is the Sun? Gleaning the Solar Power"
Print on paper
A1 (59.4 x 84 cm)
2007
Site-specific artwork created for "Weather Report: Art and Climate Change," Boulder, Colorado

Ellen K. Levy
"2020 Vision 1," 2007
"2020 Vision 2," 2005
"Sea Currents," 2005
Digital prints, enamel, acrylic
Each 80" x 24"
Courtesy of Michael Steinberg Fine Art, New York

Iñigo Manglano-Ovalle
"Iceberg (r11i01)"
Photographic documentation of installation at the Art Institute of Chicago, 2005
Original artwork: suspended anodized aluminum tubes, rapid prototyped ABS plastic, nylon
25' x 20' x 18'
2004
Courtesy of the artist and Max Protetch Gallery, New York

"Iceberg (r390i01), No. 1"
Photographic documentation of installation at the Tate, England, 2004
Original artwork: titanium alloy, fiberglass
79" x 38" x 58"
2004
Courtesy of the artist and Max Protetch Gallery, New York

"Iceberg (r11i01) Beyond the Irish Sea"
Photographic documentation of installation in Liverpool, England, 2005
Original artwork: suspended anodized aluminum tubes, rapid prototyped ABS plastic, nylon, shipping container, weathervane
Variable
2005
Courtesy of the artist and Max Protetch Gallery, New York

Patrick Marold
"Windmill Project: Iceland"
Photographic documentation of installation in Iceland
Original artwork: 200 windmills about 4' high (tin fan blades, generators, incandescent bulbs, wood mounts)
Digital print
2000-01
Courtesy of Artyard and Peggy Mangold, Denver, Colorado

"Windmill Project: Vail, Colorado"
Photographic documentation of installation in Vail, Colorado
Original artwork: 2700 windmills about 8' high (aluminum wind cups, generators, electrical components and LED bulbs, mounted in polycarbonate tubes)
Digital print
2007-ongoing
Courtesy of Artyard and Peggy Mangold, Denver, Colorado

Natasha Mayers
From "Whitefield Welcomes Global Warming/Whitefield Concerned Citizens Urge Action on Global Warming"

"Hot new fashions from L.L.Bean winter collection"

"Trumpeter with sign at beginning of the parade float"

"World on fire sign introduces second half of the parade, Whitefield concerned citizens urging action on global warming Bush fiddling while the earth is burning in the foreground and Maine tropical fruit grower (end of first half of those in favor of global warming)"

"Ice cream vendor and sunscreen, both happy about global warming"

"2006 4th of July Parade, Whitefield, Maine: Whitefield Welcomes Global Warming and Whitefield Concerned Citizens Urge Action on Global Warming" (group photo)
4th of July Parade, Whitefield, Maine

"Bush Fiddling While World Burns: Maine Fiddlers Protest Bush inaction on global warming," Kennebunkport, July 3, 2006
4th of July Parade, Kennebunkport, Maine
Color Photographs
2006

Jane McMahan
"Arapaho Glacier: What Goes Around Comes Around"
Arapaho Glacier ice, glass, steel, refrigeration equipment, solar panels, batteries, aluminum screen, pump, hose, Boulder Creek water
Ice box 72" x 32" x 32", solar frame 6' x 10' x10'
2007
Site-specific artwork created for "Weather Report: Art and Climate Change," Boulder, Colorado

Mary Miss
"Connect the Dots: Mapping the High Water, Hazards and History of Boulder Creek"
Mixed Media
2007
Site-specific artwork created for "Weather Report: Art and Climate Change," Boulder, Colorado

The artist would like to recognize the following sources: Boulder Area Sustainability; Information Network; Boulder Creek Flood Notebook; Boulder County Government; Boulder Magazine; Daily Camera; Carnegie Branch of the Boulder Public Library; City of Boulder; Denver Public Library; Denver Regional Council of Governments; Dr. Spenser Havlick; ESRI – Ester Worker, Director of Education Programs; Gilbert F. White Papers on Boulder Creek Floodplain History; South Boulder Creek Mapping Study; University of Colorado at Boulder; the US Geological Survey; the Colorado Division of Water Resources; and the National Oceanic & Atmospheric Administration.

Joan Myers
"Gold Harbor, S. Georgia"
Digital print using Cone all-black inks on Hahnemühle paper
12.5" x 34"
2001
Courtesy of the artist and Andrew Smith Gallery

"Barne Glacier, Antarctica"
Digital print using Cone all-black inks on Hahnemühle paper
12.5" x 34"
2002
Courtesy of the artist and Andrew Smith Gallery

"Grounded Iceberg, Antarctica"
Digital print using Epson K3 inks on Hahnemühle paper
12" x 34"
2004
Courtesy of the artist and Andrew Smith Gallery

"Beacon Valley, Antarctica"
Digital print using Epson K3 inks on Hahnemühle paper
22" x 14.5"
2002
Courtesy of the artist and Andrew Smith Gallery

"Adélie, Beaufort Island"
Digital print using Epson K3 inks on Hahnemühle paper
14.5" x 22"
2003
Courtesy of the artist and Andrew Smith Gallery

"Iceberg, Crystal Sound"
Digital print using Epson K3 inks on Hahnemühle paper
30" x 20"
2001
Courtesy of the artist and Andrew Smith Gallery

"Sea Ice, Antarctica"
Digital print using Cone all-black inks on Hahnemühle paper
22" x 14.5"
2002
Courtesy of the artist and Andrew Smith Gallery

"Ice Crack, Antarctica"
Digital print using Cone all-black inks on Hahnemühle paper
34" x 14"
2002
Courtesy of the artist and Andrew Smith Gallery

Beverly Naidus
From the series "Brought to You by 'Men Who Plan Beyond Tomorrow'"
"Generally Alarming," 2006
"Doom Plastics," 2006
"Wastinghouse," 2007
"Weather Change," 1998
"Monstrous," 2007
"Crashing Halt," 2003
"Union Ecocide," 2006
"Deserts Will Bloom through Atomic Power," 2004
Digital monoprints
8.5" x 11"

Chrissie Orr
"REMNANTS: Earth Drawing 4"
Recycled corten steel, mulch, compost, soil, clay, rock, pebbles, seeds, turmeric, found natural materials
2007
Site-specific artwork created for "Weather Report: Art and Climate Change," Boulder, Colorado

"Earth Drawing" (1, 2, 3, 4)
Watercolor paper, India ink, walnut ink
#1-3, 15" x 22.5" and #4, 15" x 45"
2007

Andrea Polli
"N."
Sound and video projection of computer-based work
2003-2006

From the series "Airlight"
"Airlight SoCal," 2007
"Airlight Taipei," 2006
"Airlight Boulder," 2007
Real-time networked computer systems on video wall
Displayed for the "Weather Report: Art and Climate Change" exhibition on the campus of the University of Colorado at Boulder in the Alliance for Technology, Learning and Society Institute, or ATLAS.

The Airlight *series downloads and translates daily amounts of O_3 (ozone), NO_2 (nitrogen dioxide), CO (carbon monoxide) and other pollutants in the atmosphere into a changing rhythmic soundscape. Live traffic webcam images from the locations are blurred based on pollutant levels. Webcams are utilized for* Airlight *from the California and Colorado Departments of Transportation and the Traffic Engineering Office of Taipei City. Daily amounts of O_3 and NO_2 for Southern California updated hourly provided by the South Coast Air Quality Management District AQMD www.aqmd.gov. Thanks to Kevin Durkee and the South Coast AQMD for assistance in retrieving the data. Hourly EPA data for Taipei formatted by Dr. Chung-Ming Liu, Director of the Global Change Research Center and Professor of the Department of Atmospheric Sciences at National Taiwan University. Hourly air quality data for Colorado provided by the AIRNow Program and The AQI and VSI Air Quality Reporting Systems of The Colorado Department of Public Health and Environment with the scientific collaboration of Paulette Middleton. Thanks to the AIRNow Data Management Center for assistance in retrieving the data.*

Marjetica Potrč
"Boulder in a Time of Global Warming"
Ink on paper
8 drawings, each 8.3" x 11.7"
2007
Courtesy the artist and Max Protetch Gallery

Aviva Rahmani
"Trigger Points, Tipping Points"
Digital prints from satellite imagery and live mixed virtual media during collaborative conversations between the artist and Dr. Jim White.
Panels presented with DVD of captured recordings.
Permission pending (for Google Earth Pro satellite mapping).
Grid of 12 digital prints, each 13" x 19" and DVD presentation
2007

Rapid Response
"Global Warning"
Photographic representation of installation at Nikolai Fine Art, New York, New York
1999
Photo credit: Julia Fischer

Buster Simpson
"Stretched Shirts"
Photographic representation of performance at Mount Hood, Oregon
Digital print
13" x 69"
2002

Kristine Smock
"Our Lady of Transformation and Renewable Energy"
Steel, recycled materials, solar panels, soil
8' x 3' x 3'
2007
Site-specific artwork created for "Weather Report: Art and Climate Change," Boulder, Colorado

Joel Sternfeld
"When It Changed"
Portraits Made at the United Nations Framework Convention on Climate Change, Montreal, Canada, November/December, 2005
Pigment prints
Overall 141.675" x 59", made up of a grid of 15 portraits, each 16.166" x 24.25"
2005
Courtesy of the artist and Luhring Augustine, New York

Mierle Laderman Ukeles with Meg Duguid
"Dear Children,
Can you forgive us?
And by the way...."
Dimensions of each individual artwork: 8.5" x 11" with colored letters and a half empty page to be completed by others.
Up to 1000 total pages
2007
Site-specific artwork created for "Weather Report: Art and Climate Change," Boulder, Colorado

Melanie Walker and George Peters
"Coal Warm Memorial"
Wind installation
2007
Site-specific artwork created for "Weather Report: Art and Climate Change," Boulder, Colorado

Ruth Wallen
"Preserving Paradise"
Digital photographs and mixed media
44 postcards 4" x 6", 2 wall panels 13" x 19", postcard stand 68" x 11.5" x 36"
2007
Photo Credit: Ruth Wallen

Sherry Wiggins
"Carbon Portraits"
3 billboards; digital prints mounted on plywood
Each 3' x 6'
2007
Site-specific artwork created for "Weather Report: Art and Climate Change," Boulder, Colorado

The Yes Men
"Exxon Vivoleum"
Video
2007

"Ice Age"
Video
2007

Shai Zakai
"Forest Tunes—The Library" (a, b)
Photographic representation of installation
Original artwork: multi-media installation, wood, organic materials, video art, photographs, sound, texts
Color photograph
20" x 13"
1995-2007
a. 120 black boxes contain leaves and seeds from all over the world as well as texts. They are collected and written by the artist, over a period of nine years, and the project is still ongoing.
Installation includes a 3.5 minutes video-art Genus Loci.
b. 770/ Box no. 103/oct.2003/ Personal-Global Diary/Kibbutz Maagan Michael/Farewell
1995-ongoing

"Adaptation & Mitigation—Learning from the Ancients"
Still images from video
Each 20" x 13"
2006-7

"Erosion" from "Genus Loci"
Photographic representation of site-specific public art installation in Geum-gang Nature Art Biennale in Korea
Original artwork: fourteen metal and recycled wood "Landscape Units," seven outdoor and seven indoor
Color photograph
20" x 13"
2004